UNDERSTANDING CLIMATE CHANGE IMPACTS ON CROP PRODUCTIVITY AND WATER BALANCE

UNDERSTANDING CLIMATE CHANGE IMPACTS ON CROP PRODUCTIVITY AND WATER BALANCE

S.K. JALOTA

B.B. VASHISHT

SANDEEP SHARMA

SAMANPREET KAUR

Academic Press is an imprint of Elsevier
125 London Wall, London EC2Y 5AS, United Kingdom
525 B Street, Suite 1800, San Diego, CA 92101-4495, United States
50 Hampshire Street, 5th Floor, Cambridge, MA 02139, United States
The Boulevard, Langford Lane, Kidlington, Oxford OX5 1GB, United Kingdom

Notices
Knowledge and best practice in this field are constantly changing. As new research and
experience broaden our understanding, changes in research methods, professional practices,
or medical treatment may become necessary.

Practitioners and researchers must always rely on their own experience and knowledge
in evaluating and using any information, methods, compounds, or experiments described
herein. In using such information or methods they should be mindful of their own safety
and the safety of others, including parties for whom they have a professional responsibility.

To the fullest extent of the law, neither the Publisher nor the authors, contributors, or editors,
assume any liability for any injury and/or damage to persons or property as a matter of
products liability, negligence or otherwise, or from any use or operation of any methods,
products, instructions, or ideas contained in the material herein.

Library of Congress Cataloging-in-Publication Data
A catalog record for this book is available from the Library of Congress

British Library Cataloguing-in-Publication Data
A catalogue record for this book is available from the British Library

ISBN: 978-0-12-809520-1

For information on all Academic Press publications
visit our website at https://www.elsevier.com/books-and-journals

Working together
to grow libraries in
developing countries

www.elsevier.com • www.bookaid.org

Publisher: Candice Janco
Acquisition Editor: Laura S Kelleher
Editorial Project Manager: Emily Thomson
Production Project Manager: Bharatwaj Varatharajan
Cover Designer: Mark Rogers

Typeset by SPi Global, India

CONTENTS

ABOUT THE AUTHORS

Name	S.K. Jalota
Qualification	PhD
Affiliation details	42-year service in the Department of Soil Science, 20 years as professor or equivalent at Punjab Agricultural University, Ludhiana-141004, India
Contact	Email: jalotask03@yahoo.com and jalota50@pau.edu Mobile phone: +91-9780445307
Specialization	Soil physics, water management, and climate change
Foreign visits	Texas A&M University, Canyon and College Station; Blackland Research and Extension Center, Temple; and Florida University, Gainesville, United States
Recognitions	Fellow of the National Academy of Agricultural Sciences (NAAS), India
	Recognition award from NAAS, India
	Jawaharlal Nehru Award for best PhD dissertation from the Indian Council of Agricultural Research, New Delhi, India
	Best book award from the Punjab Agricultural University, Ludhiana, India
Published work	3 authored books, 72 research papers (referred), 2 reviews, and 5 bulletins

Name	B.B. Vashisht
Qualification	PhD
Affiliation details	16-year service in the Department of Soil Science, Punjab Agricultural University, Ludhiana-141004, India
Contact	Email: bharatpau@pau.edu Mobile phone: +91-9463035049
Specialization	Specialization in modeling climate change impact
Foreign training	In the University of Minnesota, United States
Published work	25 research papers

Name	Sandeep Sharma
Qualification	PhD
Affiliation details	7-year service in the Department of Soil Science, Punjab Agricultural University, Ludhiana-141004, India
Contact	Email: sandyagro@pau.edu
	Mobile phone: +91-7888415145
Specialization	Specialization in microbiological aspects on GHG emission
Published work	35 research papers

Name	Samanpreet Kaur
Qualification	PhD
Affiliation details	13-year service in the Department of Soil and Water Engineering, Punjab Agricultural University, Ludhiana-141004, India
Contact	Email: samanpreet@pau.edu
	Mobile phone: +91-9501114235
Specialization	Specialization in climate change and groundwater
Foreign training	In Ohio State University, United States
Recognition	Got the Jawaharlal Nehru Award for best PhD thesis in year 2015 on climate change and groundwater
Published work	35 research papers

PREFACE

At present, climate change is a reality. In many parts of the world, increased temperature has already reduced the crop productivity; increased CO_2 has increased water-use efficiency by enhancing photosynthesis and reducing transpiration in plants; changed precipitation has influenced the status and quality of groundwater by altering snow melt, drought and floods, rise in sea level, and intrusion of salt water in the coastal areas from the ocean. More or less, all the existing climate change scenarios indicate higher magnitude of climate change in the future than the present, which may further exacerbate the adverse effect on crop productivity, available water resources, and food security for the burgeoning population. It warrants proper understanding of the fundamentals of climate variability and change; their effects on physical, chemical, and biological processes in soil–plant–atmosphere continuum; crop productivity; and water balance of root zone and groundwater. Although abundant advancement on these aspects has been made, yet, the information is scattered hither and thither separately and needs synthesis to (i) make that accessible at one source and (ii) understand the interactions of changed climate variables with different processes in soil, plant, and atmosphere in a better way. Such synthesis can help the climate change personnel to comprehend the role of atmosphere, soil, and plants in greenhouse gas (GHG) emission and design adaptation technologies and mitigation strategies to combat climate change impact and secure food and water resources for the upcoming. In view of that, a book entitled *Understanding Climate Change Impacts on Crop Productivity and Water Balance* comprising of five chapters has been written.

Chapter 1 highlights the sources, processes, and factors controlling formation and escape of main greenhouse gases (GHGs) such as nitrous oxide, carbon dioxide, and methane and methods of their measurements. The role of atmospheric, plant, and soil parameters and management interventions in GHG emission and its feedback on climate change are also elucidated besides the potential global warming, radiative forcing, and lifetime of GHGs in the atmosphere. Chapter 2 illustrates global climate models, their downscaling (statistical and dynamic), different emission scenarios from Special Report on Emission Scenarios (SRES) to Representative Concentration Pathways (RCPs) and uncertainties for better understanding of climate change projections, ways to minimize bias in the projected climate data by bias correction methods, and uncertainties by performing ensemble simulations to have a more robust estimate of the climate change. Chapter 3 covers the fundamentals of direct and interactive effects of climate variables in the changing climate scenario on soil environment (carbon pools, microbial population and diversity, nutrient availability, and soil water) and processes in plant (photosynthesis, respiration, transpiration, crop duration, and phenology) to comprehend their impact on yield and quality of agricultural crops, field water balance components, and water-use efficiency. Climate change effect as modified by photosynthetic pathway (C3 and C4) and N-fixing capability (legumes and nonlegumes) of the crops, water regime, and nitrogen level in the soil is also discussed. Chapter 4 comprises the impact of climate change, land use and land cover, vegetation and geology of the aquifer on recharge and discharge, fluxes across boundaries, and consequently change in groundwater status. The emission of gases due to energy expended for pumping the groundwater; groundwater modeling by coupling climate, soil-water-plant, and groundwater models through geographic information system (GIS); and management interventions to protect the groundwater resources from the impact of climate change are also discussed. Chapter 5 covers the basics of adaptation technologies to combat the climate change impact on crop production and mitigation strategies to minimize GHG (nitrous oxide, carbon dioxide, and methane) emission in agriculture.

The book written systematically in lucid and friendly way will be of great use to both under- and postgraduate students and researchers in various universities and other institutions in disciplines of climate change, agrometeorology, soil science, soil and water engineering, and environmental science.

S.K. Jalota
B.B. Vashisht
Sandeep Sharma
Samanpreet Kaur

CHAPTER ONE

Emission of Greenhouse Gases and Their Warming Effect

Contents

1.1 INTRODUCTION

The atmospheric concentration of greenhouse gases (GHGs) such as nitrous oxide (N_2O), nitric oxide (NO), carbon dioxide (CO_2), methane (CH_4), ozone (O_3), and halocarbons is increasing significantly over time as a result of biotic (plants, animals, fungi, bacteria, etc.) and anthropogenic (industry, mining, transportation, construction, habitations, deforestation, etc.) activities. These gases are going to increase further in the future with rise in population, crop production, and changing consumption patterns for food including increasing demand for ruminant meats and are likely to bring major changes in global environment. The main sources contributing to GHGs are combustion of fossil fuels and industrial processes. Agricultural sector and land use also contribute noticeably.

Emission of different GHGs and their contribution to the total emission at global level varies with the land use. At global level, N_2O, CO_2, and CH_4 contribute 8%, 77%, and 15%, respectively, to the total emission. The agriculture sector contributes 32% of total global emissions, of which 6%, 18%, and 8% are by N_2O, CO_2, and CH_4, respectively (De la Chesnaye, Delhotal, DeAngelo, Ottinger Schaefer, & Godwin, 2006). Among the different sources of N_2O emission, 67% is natural, and 33% is anthropogenic (Davidson & David, 2014). Out of the total anthropogenic sources of N_2O emission, 66% is from agricultural lands; 11% from biomass burning; 15% by industry, energy, and transport; and the rest from other sources (Fig. 1.1). Likewise, 65% of N_2O emission from soil (Mosier, 1998) and 12%, 11%, and 7% from biomass burning, rice agriculture, and manure management, respectively, are documented (US-EPA, 2006). The agriculture sector accounts for about 10%–12% of global anthropogenic GHG emissions, of which 54% is as N_2O (Smith et al., 2007).

The emission of CO_2 is mainly from burning of fossil fuel, industry, and transport, which constitutes about 70% of the total emissions (GOI, 2015).

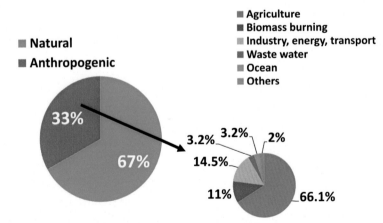

Fig. 1.1 N_2O emission inventory.

From agricultural lands, CO_2 is emitted from burning of biomass and fossil fuels, tillage and soil disturbance, decomposition of organic matter (OM), deforestation, draining of wetlands, uncontrolled grazing, manufacturing of fertilizers and pesticides, etc. But it is <1% of global anthropogenic CO_2 emission.

The atmospheric CH_4 instigates from natural (wetlands, oceans, forests, wildfires, termites, geologic sources, and gas hydrates) and anthropogenic (agriculture, energy production and transmission, and waste and landfills) sources. In 2004, the anthropogenic contributions to CH_4 emission by agriculture, energy production and transmission, and waste and landfills were 43%, 36%, and 18%, respectively (IEA, 2006). The emission of CH_4 in agricultural sector is from rice fields and enteric fermentation in ruminant, which constitute 30% and 18%, respectively, of the total anthropogenic emissions (Bodelier, Roslev, Henckel, & Frenzel, 2000). The CH_4 emission of 6% from manure management, 10% from rice, and 40% from enteric fermentation has been reported as well (Kasterine & Vanzetti, 2010).

In the future, these capricious contributions are going to change at temporal and spatial scales depending upon the adaptation technologies and mitigation strategies followed under the changed land use at regional scale to manage the amount and type of increasing food demand for the increased population. There may be more emission of GHGs from (i) increased demand of livestock products and resulting intensification of agriculture, especially in unexploited areas; (ii) more use of N fertilizers owing to change in land use, that is, rapid conversion of forests to croplands and increasing cattle population; and (iii) non–CO_2 manure management and N_2O emission from soils. However, in Western Europe, emission of GHGs has reduced and projected to decrease in 2020 due to adoption of climate and other environmental policies.

From agriculture sector, the emissions of N_2O, CO_2, and CH_4 together account for approximately one-fifth of the annual increase in radiative forcing of climate change (Cole et al., 1997). However, the contribution of an individual GHG depends upon its radiative forcing (i.e., *change in the balance between incoming and outgoing radiation to the atmosphere*), lifetime in the atmosphere, and consequential total emission, which decide the warming impact. The emitted GHGs in the atmosphere keep the earth warm by acting as a shield around the earth and maintaining temperature by absorbing the reflected infrared radiations (long wave), known as *greenhouse effect*. This effect is being continuously enhanced by increasing level of GHGs in the atmosphere and results in rise of temperature and acid deposition at global level. Among different GHGs, N_2O not only causes warming but also is responsible for destructing the stratospheric ozone too.

For understanding the emission of GHGs and their warming effect, it is prerequisite to have knowledge of sources, processes, and factors influencing formation and emission of GHGs drawn in the nitrogen (N), carbon (C), and methane (CH_4) cycles in soil-plant-atmosphere continuum (SPAC), lifetime of GHGs in the atmosphere, and radiative forcing. In this chapter, all the aspects that include the role of atmospheric, plant, and soil parameters and management interventions and feedback of climate change on GHGs are also discussed besides the potential global warming, radiative forcing, and lifetime of GHGs in the atmosphere. Measurements by different methods and estimation by empirical relations and mechanistic models are also described.

1.2 NITROGEN CYCLE

The sources of N are atmosphere, soil, and water on the earth. A huge amount of N (78.3% by volume) exists in these sources, but more than 99% of this is in the form of dinitrogen gas (N_2). The N_2 is unavailable to almost all living organisms because of the triple covalent bond between N atoms ($N \equiv N$), which does not allow the N_2 molecules easily enter into chemical reactions. Once the bond is broken, N_2 becomes reactive in the form of organic (urea and amines) and inorganic N compounds of reduced (NH_3 and NH_4^+) and oxidized (N_2O, NO_2, and HNO_3) forms. In the soil, N exists in various forms having different oxidative states (Table 1.1).

Table 1.1 Main Forms of Nitrogen in Soil and Their Oxidation States

Name	Chemical Formula	Oxidative State
Nitrate	NO_3^-	+5
Nitrogen dioxide (gas)	NO_2	+4
Nitrite	NO_2^-	+3
Nitric oxide (gas)	NO	+2
Nitrous oxide (gas)	N_2O	+1
Dinitrogen (gas)	N_2	0
Ammonia (gas)	NH_3	−3
Ammonium (gas)	NH_4^+	−3
Organic N	RNH_3	−3

Some other forms of N (proteins, peptides, labile N, hydrolysable unknown N forms, acid insoluble N associated with aromatics, amino sugars N, and nucleic acid like purines and pyrimidine derivatives) are also liberated from soil organic matter (SOM), but the quantity of those is <5% of the total. The quantity of N in soil is also controlled by the type of land use, for instance, agricultural and forest soils have 10–20 times more N than the standing vegetation. In cropped soils, N is cycled by biologically mediated redox reactions. Change in chemical speciation in these reactions often results in the movement of N from one reservoir to another in different physical locations. For example, when synthetic N fertilizer is applied to crops, a part (50%) of it is taken by the plants, and the rest is lost as nitrate (NO_3^-) leaching, erosion, and gaseous emissions. The leached NO_3^- and volatilized NH_3 are changed to N gas, N_2O, and NO that escape into the atmosphere. The N removed from the field in crop is taken by animals, which generate manure (urine and feces). In this process, about 20% of N in the manure is volatilized as ammonia (NH_3) gas. The manure (organic N) is returned in croplands to fertilize the next crop, thereby completing the N cycle. In the N cycle, there are a number of biochemical processes in soil, which transform one form of N to another through physical (nonbiological) processes and are cycled in SPAC. The biological transformations mediated by microorganisms are N mineralization and immobilization, N fixation, nitrification, and denitrification (Fig. 1.2).

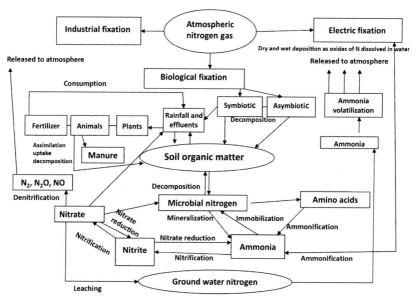

Fig. 1.2 Nitrogen cycle.

1.2.1 Biochemical Processes

Nitrogen Mineralization and Immobilization

Mineralization and immobilization are two main biogenic processes in N cycle carried by a wide array of microorganisms such as aerobes, anaerobes, fungi, and bacteria in soil. These microorganisms attack dead biomass (detritus) primarily as a source of energy and C to support their growth and convert organic forms of nutrients in detritus into soluble forms usually ammonium (NH_4^+) that can be again taken by plants and other microbes. Soil fauna also plays an important role in these processes by regulating populations of bacteria and fungi. The microorganisms themselves have a need for nutrients, especially N, to assemble proteins, nucleic acid, and other cellular components. If plant detritus is rich in N, microbial needs are met easily, and N releases or mineralization proceeds through reactions of aminization and ammonification. In aminization, freshly added organic materials like proteins and other complex compounds (amino acids, amines, amides, urea, chitin, and amino sugars) are attacked by a group of organisms, which break down the protein structure by enzymatic digestion process and release amino-N (Eq. 1.1a). In ammonification, the released amines $(R-NH_2)$ are changed to NH_3 by another group of heterotrophs (Eq. 1.1b):

$$Proteins \rightarrow R - NH_2 + CO_2 + Energy + otherproducts \qquad (1.1a)$$

$$R - NH_2 + H_2O \rightarrow NH_3 + R - OH + Energy \qquad (1.1b)$$

If plant detritus is low in N, microorganisms would scavenge additional N from their surroundings to build proteins, and there would be immobilization. The immobilization process is opposite that of the mineralization, where mineral N (NH_3 and NO_3^-) and even organic N (amino acids) are consumed by diverse groups of microorganisms to synthesize their protein and get multiplied. Thus, release of inorganic N during decomposition and demand on the soil N pool made by decomposer organisms for assimilating N into microbial tissues depend upon the C:N ratio of the substrate. Microbial N need is also affected by organism's growth efficiency. For example, fungi have less N needs and grow more efficiently on low N substrates than bacteria owing to wider C:N ratios. In general, the materials with a C:N ratio <25:1 stimulate mineralization, while those with a C:N ratio >25:1 stimulate immobilization. For instance, wheat straw (with C:N ratio 80–100:1) leads to net N immobilization, whereas legumes (with C:N ratio 12–15:1) lead to net mineralization. Actually, it is the C:N ratio that provides an opportunity to manage the synchrony of N supply to plants or to immobilize inorganic N from the soil pool so that the chances of N loss by denitrification and leaching are reduced. Thus, to manage N losses, it is imperative to manage C:N ratio to an optimum level depending upon the substrate quality and composition of the decomposer community.

Nitrogen Fixation

Nitrogen fixation is a route through which the very stable and unavailable form of N is transformed to reactive and available form by breaking the triple covalent bond between N atoms ($N \equiv N$) so that N can bond with C, hydrogen (H), and oxygen (O). In this process, a lot of energy is required. The energy to fix N occurs in three main places, that is, atmosphere, tiny microbial bacteria, and industry. In the atmosphere, N fixation occurs when the enormous energy of lightning breaks apart the N molecules allowing them to bond with oxygen in the air, forming nitrogen oxides ($NO + NO_2$) and N_2O forms. These forms of N dissolve easily into rain that is then carried down to the earth. In microbial N fixation, inert N present in atmosphere is fixed by only a very selective few bacteria and archaea (*the most primitive living single-celled organisms, similar in size to bacteria, but different in molecular organization*) through their specialized enzymes, nitrogenases. The nitrogenase enzyme has two keyholes, one for N and one for H. When the two elements fit into the keyholes, the nitrogenase enzymes break the bond between the two N molecules and connect them with H. Using tremendous amount of energy, namely, adenosine triphosphate (ATP), nitrogenase squeezes together one N molecule and three H molecules forming two weak NH_3 molecules, given away in Eq. (1.2), representing the biological N fixation:

$$N_2 + 8H^+ = 8e^- + 16ATP \rightarrow 2NH_3 + H_2 + 16ADP + 16P_i \qquad (1.2)$$

During microbial N fixation, archaea and bacteria use diverse energy sources, for example, sunlight by phototrophs, reduced inorganic elements and compounds by lithotrophs, and a glut of different organic substrates by heterotrophs. The phototrophs, lithotrophs, and heterotrophs correspond to obligate aerobes, facultative anaerobes, and obligate anaerobes, respectively. Biological N fixation with the nitrogenase enzyme complex (dinitrogenase and nitrogenase reductase) is carried out exclusively by different types of microorganisms, that is, those that (i) live symbiotically in nodules on the roots of leguminous plants; (ii) fix N in root nodules of some nonleguminous plants; (iii) live on or close to the soil surface and are photosynthetic; and (iv) live in association with plant roots, but nonsymbiotically.

Approximately $100–140\text{ Tg N year}^{-1}$ enters the biosphere by biological N fixation and during lightning discharges (Galloway, Schlesinger, Levy, Michaels, & Schnoor, 1995). In industrial N fixation, N_2 gases are forced to react under huge pressure (500 atm) and temperature (1200°C) in the presence of finely divided Fe_2O_3 catalyst to break the triple covalent bond and form ammonia. The ammonia so produced is directly used as fertilizer in soil. Sometimes burning of fossil fuels also makes the NH_3 so hot that the N_2 molecules break apart and add N to the atmosphere.

Nitrification
Nitrification is the microbial oxidation of reduced forms of N to less reduced forms, principally NO_2^- and NO_3^-. Under favorable environmental conditions (oxic, pH near neutral), free NH_4^+ ions rarely persist in soils and rapidly oxidize to NO_3^- ions by chemoautotrophic nitrifier, for which NH_4^+ is the only source of energy. The reaction typically occurs in two stages that are carried out by a few genera of bacteria, termed the "nitroso" genera, *Nitrosomonas, Nitrosovibrio, Nitrosolobus,* and *Nitrosococcus,* that carry out oxidation of NH_4^+ to NO_2^- (Eq. 1.3a) and the "nitro" bacteria, *Nitrobacter, Nitrospira, Nitrococcus,* and *Nitrospina,* that carry out the oxidation of NO_2^- to NO_3^- (Eq. 1.3b). But because of the high affinity of the organisms, carrying out the second stage for nitrite (NO_2^-), the intermediates rarely accumulate:

$$\text{Primary nitrification}$$
$$2NH_4^+ + 3O_2 \rightarrow 2NO_2^- + 2H_2O + 4H^+ + \text{energy} \qquad (1.3a)$$
$$\text{Nitroso} - \text{bacteria}$$

$$\text{Secondary nitrification}$$
$$2NO_2^- + O_2 \rightarrow 2NO_3^- + \text{energy} \qquad (1.3b)$$
$$\text{Nitro} - \text{bacteria}$$

There is also increasing evidence that fungi may oxidize NH_4^+, but this is believed to be restricted to acidic forest soils. The NO_3^- formed during nitrification becomes the starting point for denitrification, so nitrification can promote denitrification even though the two processes are supported by contrasting environmental conditions (denitrification is an anoxic process, whereas nitrification is an oxic process).

Factors Affecting Nitrification
Nitrification being more sensitive to soil environment as compared with N mineralization necessitates understanding the effects of different physical,

chemical, and biological properties of soil and cultural practices on nitrification, which are discussed below:

Soil reaction: The nitrifying microorganisms are sensitive to soil reaction (pH). Optimum nitrification occurs at pH 6.6–8.5 because supply of Ca^{++} and $H_2PO_4^-$ and adequate level of essential elements in this range make the soil environment favorable for microflora related to nitrification. At pH >8.5, the secondary nitrification, transformation of NO_2^- to NO_3^-, is inhibited by NH_4^+ as it may rise to toxic level for nitrobacterial growth. Such conditions facilitate NO_2^- accumulation. At pH <4.5, nitrification becomes negligible due to toxic effects of aluminum or manganese, which restrict the major exchangeable cations. However, some nitrifier fungi and heterotrophic bacteria nitrify organic N directly to NO_3^- without the formation of NH_4^+ as they are able to tolerate acidic pH in soil through slime production or localized cluster formation with soil aggregates.

Soil aeration: Nitrifying organisms are obligate aerobes and require oxygen (O_2) to form NO_2^- or NO_3^- ions; hence, scanty of O_2 in soil diminishes their activity. Nitrification is optimum when aeration porosity is 50%–60% of the total porosity. In aggregated soils, nitrification is confined mostly to the outer part of an aggregate because of rapid diffusion of O_2 in the air.

Soil moisture: Nitrification proceeds readily from −0.3 to −1.6 MPa soil moisture potential. Above −0.6 MPa, dehydration impacts become critical. Nitrification is limited at waterlogged or saturated conditions. The effect of dehydration (water stress) on nitrification is more by autotrophic bacterial nitrifier than fungal nitrifier. Nitrification mostly occurs during dry spell and produces NO_3^-.

Soil temperature: The optimum range for NO_2^- oxidation lies between 25°C and 35°C. The process of NO_2^- oxidation is slow when soils are cool or warm and virtually ceases at temperature <5°C and >40°C. Despite this, nitrification proceeds at even 0°C in forest soils because of the predominance of low-temperature-adaptive nitrifying soil fungi. Seasonal temperature fluctuations in soil also influence the nitrification. Nitrification is greatest in spring and fall and is limited under low temperature in temperate soils.

Soil organic matter: Nitrifiers being autotrophs do not require OM as either a C or an energy source. However, in case of OM decomposition though the supplies of NH_4^+ and O_2 to nitrifier may temporarily get limited due to inorganic N and O_2 requirement, nitrification will be enhanced because of the survival of both nitrite and ammonium oxidizers in low-oxygen environments (Ward, 2011).

Population of nitrifying organisms: Soils may differ in their ability to nitrify NH_4^+ even under similar conditions of temperature, moisture, and level of added NH_4^+ because of the variation in population and group of nitrifying organisms. There are three groups of nitrifier. First group is the NH_3 oxidizers, which are obligate autotrophs and can grow only by fixing their own CO_2 using the Calvin cycle; their source of energy is only NH_3; and they oxidize NH_3 to NO_2^-. The second group is NH_3-oxidizing archaea, not bacteria. They predominantly are autotrophic and fix CO_2 using the 3-hydroxypropionate/4-hydroxy butyrate pathway rather than the Calvin cycle. They oxidize NH_3 to NO_2^- and produce N_2O and NO_2^-. The third group is of bacteria, members of the Planctomycetes phylum, capable of oxidizing NH_3 using NO_2^- instead of oxygen and producing N_2 instead of NO_2^-.

Cultural practices and crop type: Practice of pesticide application generally reduces nitrification due to their toxic effect on soil microbes, while that of tillage promotes nitrification by improving soil aeration. Plowed grasslands usually result in ample NO_3^- production. Certain crops such as *Brachiaria humidicola* and *Azadirachta indica* have inhibitory effect on nitrification through root exudates because of their chemical composition, brachialactone, containing cyclic diterpene with a unique 5-8-5-membered ring system and a c-lactone ring, which contributes 60%–90% to inhibitory activity.

Denitrification

Denitrification is the reduction of NO_3^- to N gases (NO, N_2O, and N_2) that occurs under anoxic conditions. The reaction proceeds in four steps via a series of intermediate catalyses by separate enzymes and a wide range of heterotrophic bacteria, which use N species as a respiratory electron acceptor in place of O_2. As all the steps of denitrification (Eq. 1.4) are not induced simultaneously and all denitrifier are not capable to complete the full reduction pathway from NO_3^- to N_2, therefore, some intermediates (N_2O and NO) can accumulate:

$$NO_3^- \rightarrow NO_2^- \rightarrow NO \rightarrow N_2O \rightarrow N_2 \qquad (1.4)$$

$$\underset{\text{Nitrate}}{} \quad \underset{\text{Nitrite oxide}}{} \quad \underset{\text{Nitric oxide}}{} \quad \underset{\text{Nitrous oxide}}{} \quad \underset{\text{Dinitrogen}}{}$$

In soil, most culturable denitrifiers are facultative anaerobes, that is, *Pseudomonas* and *Alcaligenes* and to a lesser extent *Bacillus*, *Agrobacterium*, and

Flavibacterium. In aerobic rice *Pseudomonas, Thiobacillus, Bacillus, Alcaligenes, Flavobacterium*, etc. cause denitrification and generate N_2O. In addition to biological denitrification, some nonbiological denitrification in subsoil called chemodenitrification also occurs and emits N_2O from agricultural soils, manure, and soilborne N, especially in fallow years, legumes, plant residues, and compost. However, emission from such sources is not significant (Granli & Bockman, 1994).

Factors Affecting Denitrification

The magnitude and rate of denitrification are strongly affected by several soil and environmental factors. The important determinants of the denitrification rate in soil are the following:

Level and form of inorganic nitrogen: Concentration of NO_3^- in soil affects the rate of denitrification and exerts a strong influence on the ratio of N_2O to N_2 gases released. At higher NO_3^- level in soil, there is an incomplete denitrification, and it results in higher $N_2O:N_2$ ratio due to suppression of nitrous oxide reductase activity, the enzyme responsible for microbial conversion of N_2O to N_2. At NO_3^- level of $>20\ \mu g\ N\ mL^{-1}$, denitrification is independent of the amount of NO_3^- present. At low NO_3^- levels, denitrification will be limited by the rate of NO_3^- diffusion to its site, and predominant emitted gas is N_2O or N_2.

Nature, amount, and solubility of carbon and decomposability of organic matter: Denitrification process in soil is increased with increasing the amount of total, water-soluble, and readily decomposable SOM content as most of the denitrification is carried out by heterotrophic bacteria, which show strong dependence on C availability. The available C substrates supply electrons, and decomposition of OM produces CO_2 and reduces O_2 supply, thus increasing demand for NO_3^- as electron acceptor during microbial growth. This process is more pronounced in freshly added crop residues, as it supplies more C and better reducing environment. By addition of organic materials in the soil, total amount of N_2O produced is enhanced; however, ratio of $N_2O:N_2$ from denitrification decreases with available C supply.

Soil reaction: Most denitrifiers show growth optima at near neutrality (pH 6.0–8.0). Denitrification is negligible in soils having pH <5.0 and absent below pH 4.0 due to inhibition of N_2O reduction. Soil acidity (low pH) decreases the formation of N_2O during denitrification by (i) lowering the decomposition rate of SOM, hence reducing the availability of N substrate for N_2O production; (ii) inhibiting N_2O reductase with the result that denitrification yields more N_2O than N_2; (iii) reducing the availability of molybdenum that in turn may reduce the synthesis of NO_3 reductase, a molybdoprotein enzyme; and (iv) imposing toxicity effects of aluminum or manganese. Apart from low pH, high salinity also decreases denitrification.

Soil temperature: Soil temperature influences denitrification; however, the rate of denitrification increases rapidly from 2°C to 35°C with optimum temperature of 25°C–35°C. At higher temperature (>50°C), denitrification is decreased as thermophilic nitrate respirers change NO_2^- to N_2 gas by chemodenitrification.

Soil oxygen status: Denitrification losses are increased when O_2 supply is too low. In fact, low O_2 inhibits denitrifying enzyme synthesis plus the electron flow to denitrifying enzyme and represses the all-nitrogen oxide reductases; however, nitrate reductase is nonsignificantly affected. The denitrification losses do not become appreciable until O_2 level is drastically reduced to concentration of <10% or <0.2 mg L^{-1} of O_2 dissolved in the solution (Firestone, Firestone, & Tiedje, 1980).

Soil water: Denitrification increases with soil saturation and waterlogging. In most of the soils, the process of denitrification occurs mainly following P_{cp} as soil pores become water saturated and the diffusion of O_2 in microsites is slowed drastically. Typically, denitrification starts to occur at water-filled pore space of $\geq 60\%$.

Soil texture and structure: Denitrification is usually more in finer-textured soils than coarser ones because of more (i) availability of surface area for microbial attachment; (ii) capacity of soils to retain moisture; and (iii) supply of C and N. Denitrification is less in well-structured soils as its porosity is more than that of the poorly structured.

The presence of plants: In cropped soils as compared with fallow, denitrification rates are higher because plants release readily available C in the form of root exudates and sloughed-off roots. Increased readily available C stimulates high microbial population, which consumes O_2 near the rhizosphere region and makes the soil conditions more suitable for denitrification. That is why it is assumed that denitrification is more in the rhizosphere than nonrhizosphere region. At the same time, denitrification process is restricted by plants through uptake of NO_3^- and NH_4^+ (Box 1.1).

Box 1.1

In the soil, N_2O is produced by at least three microbial-mediated mechanisms:

1. Nitrification, utilizing as an alternative electron acceptor, thereby reducing it to N_2O.
2. Dissimilatory nitrate reduction (denitrification) *refers to the anaerobic transformation of nitrate to nitrite and then to ammonium* and is probably the main source of N_2O in soil.
3. Assimilatory nitrate reduction is *defined as the production of N gas (mainly N_2O) under aerobic environments*; it is of minor importance in soils ($<6\%$ of total nitrate reduction) because it is inhibited by low concentration of ammonium or soluble organic N present in soil.
4. Recently discovered, anaerobic ammonium oxidation (anammox) in which ammonium and nitrite are converted to N_2 or N_2O.

1.2.2 Physical Processes

Volatilization

Volatilization is a process in which NH_4^+ present in fertilizers or converted readily in soil is lost as NH_3 gas to the atmosphere because of its short lifetime. The factors affecting volatilization are soil (pH and clay mineral), climate (temperature and wind), and management (type, rate, and method of fertilizer application; their synchronization with irrigation water timings; use of urease inhibitors; etc.). Ammonia volatilization is more at pH >7, but can also occur at neutral and pH <7. In alkaline pH, the NH_4^+ is converted to NH_3 gas and escapes to the atmosphere, while in acidic pH NH_4^+, ions interact with soil cation-exchange complex or may get fixed in clay lattice. Ammonia volatilization is more pronounced in soils with clay content of $<25\%$, which lack the ability to resist pH changes (i.e., having low pH buffering capacity). Temperature increases NH_3 losses on account of increased urease activity and higher proportion of NH_3 in NH_4^+/NO_3^- equilibrium. Wind increases NH_3 volatilization linearly by transport process in the air. Both high temperature and wind turbulence increase gradient of NH_3 between soil and atmosphere and stimulate its volatilization. In general, NH_3 losses increase with N rate, but the increase may be linear (in the range of lower N rates) or exponential (in the range of lower to higher N rates). Losses of NH_3 are more from (i) broadcasted fertilizers than incorporated; (ii) liquid organic material than urea than green manure; and (iii) flooded rice fields after the application of ammonical fertilizers (Denmead, Freney, & Simpson, 1979).

Leaching

Nitrogen in NO_3^- form gets leached easily owing to its high solubility and mobility in water. The magnitude of NO_3^- loss through leaching depends on the type of fertilizer applied, amount of NO_3^- present, amount and intensity of the precipitation (P_{cp}), amount of irrigation water, and hydraulic conductivity of the soil. At comparable N application rates, NO_3^- leaching followed the order compost < unfertilized < organic fertilizers < inorganic fertilizers (Tejada & Gonzalez, 2006). Relatively lower N leaching in organic fertilizers than inorganic fertilizers is because of decrease in soluble N in the soil owing to increased (i) efficiency of denitrifiers; (ii) immobilization of NO_3^-; and (iii) capture of N in the SOM. It portrays that NO_3^- pool in subsoil or groundwater could be reduced with application of organic manures to the soil as OC is the main limiting factor for denitrification in the subsoil. Leaching is also controlled by water input at soil surface by P_{cp} plus irrigation water and soil texture. High input of water enhances movement of NO_3^- to elsewhere by runoff and down in subsurface. Leaching losses are more in coarse- than fine-textured soils because of more saturated hydraulic conductivity and low clay content. Sometimes N_2O released as a result of incomplete denitrification in subsoil could also leach by drainage water.

Erosion

Nitrate-N is displaced from the soil by wind and water. During water erosion, each ton of soil displaced from field may carry 0.5–1.5 kg N with it (Lambert, Devantler, Nesip, & Penny, 1985). Though in runoff NO_3^- is removed from topsoil, that is not comparable with that of the bulk soil removal. Nitrate may also be transported off the site during wind erosion, and the loss is limited and proportional to particulate movement. Drought usually precedes wind erosion, and during drought, capillary movement of water to soil surface may enrich the soil surface with NO_3^-. Heavy and torrential P_{cp} can cause runoff, denitrification, and also leaching, especially if the soil is relatively more porous in nature.

1.2.3 Nitrous Oxide Emission

The main natural resources of N_2O emission are land and ocean, which contribute 65% and 30%, respectively (IPCC, 2001). On the land, the main source of the global anthropogenic N_2O emissions is agriculture, which contributes largely through (i) soil management activities, including the use of synthetic and organic fertilizers, production of N-fixing crops, and cultivation of soils with high OM; (ii) manure management including storage and application of livestock manure to croplands and pastures; (iii) urine and feces deposition in grazed pastures; and (iv) crop residue management (e.g., residue burning). The emissions can be at upstream of cropping system (prefarm—production of fertilizers) and downstream (at farm—use of fertilizers for crop production). Upstream emissions are generally greater as there is more consumption of energy than downstream. In the downstream, N_2O emission is of two types.

One is the direct in which N_2O is emitted from all the practices/sources that directly put in additional N to soils by the activity of microorganisms during nitrification and denitrification processes. The other type is indirect in which N_2O is emitted as a result of other on-farm N losses like mineralization of organic fertilization (animal manures), NH_3 volatilization and NO_3^- leaching and their subsequent transport to off-farm environments, and deposition elsewhere in the landscape. Indirect emissions represent 13%–17% of the direct (Garnier et al., 2009). For synthetic N fertilizer, upstream emissions constitute 50%. The remaining 50% is emitted as downstream from the soil (Tirado, Gopikrishna, Krishnan, & Smith, 2010). However, Biswas, Barton, and Carter (2008) estimated that N_2O emission in terms of CO_2 equivalent was 3.8 times in urea production than from the field to produce 1000 kg of rain-fed wheat in Australia. Surface runoff and leaching of NO_3^- into groundwater or surface water can reemerge in riparian zones, where it becomes emitted as N_2O. The N_2O emissions may also occur during biological N fixation by legumes, but that is insignificant (IPCC, 2006). Some of the N contained in agricultural products eventually ends up in sewage treatment plants following human consumption, where it can be emitted as N_2O. Apart from synthetic N fertilizers, livestock manure is also a significant source of N_2O. However, the magnitude of N_2O emission changes with storage condition (more emission in aerobic initially and anaerobic later on), storage time (more emission in longer time of storage), type of application (more emission in surface application than injected) (Saggar, Bolan, Bhandral, Hedley, & Luo, 2004), and also time since application and supplemental water additions. The N_2O is also emitted in grazed pastures from feces patches depending upon the N present in the fecal piles that has to undergo nitrification and denitrification (Saggar et al., 2004) and is influenced by the species of forage being consumed by the animal (Archibeque, Burns, & Huntington, 2001) and the N content of the forage (Archibeque, Burns, & Huntington, 2002). There may be large temporal variations in N_2O emission rates in grazed pastures due to variations associated with grazing and fertilizer application timings, uneven distribution of excretal returns of grazing animals, animal treading effect on soil, soil temperature, and water content. Globally, more than 40% of the emissions of N_2O are attributable to agricultural systems managed for grazing animals (Denman et al., 2007).

Factors Affecting Nitrous Oxide Emission

The N_2O emissions are influenced by the characteristics of fertilizer, organic carbon (OC), properties of soil, and type of crop.

Effect of fertilizer type and application rate: N_2O from fertilizers is emitted following the order of liquid organic > organic-synthetic mixtures > synthetic fertilizers > solid organic fertilizers > unfertilized. Relatively higher emission levels in liquid organic fertilizers compared with solid are due to the presence of highly mineralizable form of the N, that is, NH_4^+. The N_2O is also emitted from soil mineral N found in the form of either NH_4^+ or NO_3^- because NH_4^+ is a substrate of nitrification and NO_3^- of denitrification. As N_2O production by nitrification or denitrification depends on the N available in the soil, so, N fertilizer application emerges as an important driver of N_2O emissions. However, the amount and fastness of N_2O emission depends upon the mineral N source in the fertilizer. The N_2O fluxes are higher from ammonium nitrate than urea because of the presence of highly mineralizable form of N. From urea, N_2O fluxes are less and delayed because it has to be hydrolyzed before being available for nitrification and denitrification processes. On the whole, the net emissions are directed by the balance among production, consumption, and diffusion of N gases to the atmosphere, which depends on N fertilization rate, source of N applied, soil water content, tillage practices, and prevailing soil temperature (Firestone & Davidson, 1989). Net emissions may range from 1 to 29 g N_2O—N $ha^{-1}day^{-1}$ (Pathak, 1999) and from 0.1% to 2% of applied synthetic N fertilizers (Saggar, Luo, Giltrap, & Maddenna, 2009).

Effect of soil organic carbon (SOC) and dissolved organic carbon (DOC): It is not necessary that a substrate with high C content is biodegradable for denitrification. With DOC (organic molecules *of varied origin and composition within aquatic systems*), denitrification rate and ratio of $N_2O:N_2$ are increased due to more C and less oxygen for microbial growth, which generate anaerobic conditions necessary for denitrification. Positive correspondence between DOC and N_2O emission parameters is only during the irrigation period. Labile C sources reduce N_2O emission strongly, and the reduction is more under high-water conditions than low-water conditions (Laverman, Garnier, Mounier, & Roose-Amsaleg, 2010).

Effect of soil water. The N_2O production is favored in waterlogged soils or anoxic microsites in moist soils. As the moisture regime of the soil advances toward the drier end, emission level and emission factor are decreased accordingly. For example, emission level of 4 kg N_2O—N ha^{-1} $year^{-1}$ in rain-fed treatments was one-tenth of that in conventionally irrigated (Eduardo, Luis, Alberto, Josette, & Antonio, 2013). Low N_2O emissions in rain-fed systems could also be linked to low fertilizer application in conjunction with low water supply. Drip irrigation showed intermediate distributions in emission levels (1.2 kg N_2O—N ha^{-1} $year^{-1}$). In general, N_2O emissions increased linearly with increasing soil moisture in the range when water-filled pore space (WFPS) is from 20% to 70%. Water-filled pore space does not affect as such, but by controlling microbial processes of nitrification and denitrification. In general, nitrification is low in soils with WFPS <40%, but increased up to 55%–65%. Water-filled pore space of 60%–70% promotes denitrification due to increased water content and hindrance of aeration, which releases both N_2O and N_2. The N_2 becomes a dominant form of gaseous N loss above 80%–90% WFPS. Under natural conditions, N_2O emission is related to increase in WFPS pulses with wetting by P_{cp} because the N_2O flux is positively related to NO_3^- concentration in soil after the onset of P_{cp} and negatively in drier soil conditions.

Effect of soil temperature: N_2O flux is positively correlated to temperature. Emission of N_2O increased with increase in soil temperature from 5°C to 40°C (Blackmer, Schepers, Varvel, & Walter-Shea, 1996). Temperature affects N_2O emissions directly through microbial activity and indirectly through decreasing WFPS due to increased soil water evaporation rate. The effect of temperature on N_2O emissions is accentuated in conjunction with water and N inputs. It is contemplated that with increased temperature in combination with water and N, the optimum temperature for denitrification is increased owing to dominance of thermophilic nitrate respirers and chemodenitrification reactions.

Effect of soil reaction: The optimum pH for N_2O emission via denitrification varies with species and age of the organisms and NO_3 concentration, but most of the denitrifiers have optimum pH for growth between 6.0 and 8.0. Although the process is favored at slightly alkaline pH, it proceeds up to pH as low as 3.5. At low pH, N_2O emissions are deceased owing to inhibition of N_2O reduction. Under saline conditions, there is N_2O

accumulation from denitrification because N_2O reductase is prone to salts also, which may result in slowdown of both nitrification and denitrification.

Effect of soil texture: Soil texture influences N_2O production through its aeration capacity constituted by soil separates that control nitrification and denitrification processes. In fine-textured soils with sediments of $<50\ \mu m$, N_2O emission via denitrification is more due to more water retention and less aeration as compared with coarse-textured soils.

Effect of tillage: N_2O emissions are higher from no–tilled soils because of less aeration, which enhances denitrification. In different tillage practices, denitrification may show heterogeneous responses because of the wide range of aeration and microbial activity.

Effect of crop type: Cumulative N_2O emissions vary with crop species, the effect of which is manifested through inputs, that is, N fertilizer rate and irrigation regime, and weather parameters (temperature and P_{cp}) that control N_2O production. The highest emissions would be in intensively irrigated and fertilized summer crops.

From the above information, it can be recapitulated that from agricultural lands, N is lost in the form of NO_3^-, NH_4^+, or NH_3, which are changed to N_2O at the end. The N_2O is released by nitrification process, when base of fertilizer is NH_3 and soil is well aerated and yet moist, and by denitrification, when base of fertilizer is NO_3 and soil aeration is restricted. The emission of N_2O depends upon the proportion of both rate of denitrification and $N_2O:N_2$ ratio, controlled by nitrous oxide reductase activity. The contribution of total N_2O emissions depends mainly on the soil, which act not only as a source, but also as a sink for N_2O that is consumed by the soil microbes due to low mineral N before being escaped from the soil, depending upon the residence time and ease of diffusion of N_2O through soil regulated by soil water content (Thomson, Giannopoulos, Pretty, Baggs, & Richardson, 2012). The imbalance of source and sink has been found responsible for the steady increase in N_2O during the past 100 years (Battle et al., 1996).

1.2.4 Effect of Climate Change on Nitrous Oxide Emission

Climate change fetches major changes in CO_2, temperature, and P_{cp} parameters of the atmosphere. The changed parameters alter the soil environment (OC, temperature, moisture, and oxygen) and biogeochemical processes related to emission of GHGs. The altered soil conditions by freeze/thaw events, wet/dry cycles, and seasonal temperatures influence the N_2O

production and its release from soil to the atmosphere (Goldberg, Trewick, & Paterson, 2008). The processes by which CO_2, temperature, and P_{cp} affect emission of N_2O are the following:

Increased CO₂

Increased atmosphere CO_2 reduces emission of N_2O by (i) increasing C:N ratio in soil, which increases immobilization; (ii) changing the genera of denitrifying bacteria, that is, more growth of fungi than bacteria, which is relatively more efficient for utilizing C and reducing N_2O emission; and (iii) limiting supplies of NH_4^+ and O_2 to nitrifiers. In some cases, N_2O emission is stimulated with the elevated atmospheric CO_2: firstly, through increased root biomass, biological activity, and soil water (details are in Chapter 3), which provide favorable conditions for denitrification, and secondly, by increasing DOC (which ultimately increases denitrification rate and $N_2O:N_2$ ratio). From a metaanalysis, using 152 observations from 49 published studies, it has been concluded by Van Groenigen, Osenberg, and Hungate (2011) that increased CO_2 (ranging from 463 to 780 ppm) stimulates N_2O emissions by 18.8% from upland soils.

Increased Temperature

Both the processes of nitrification and denitrification leading to N_2O emission are temperature-sensitive. The optimum temperature range for nitrification and denitrification processes lies between 25°C and 35°C and so is for the N_2O emission (Blackmer, Bremner, & Schmidt, 1980). Temperature >35°C decreases denitrification by increasing both O_2 solubility and O_2 diffusion in water and consequently decreases N_2O emission. At temperature <5°C and >40°C, more N_2O emission is expected because the process of NO_2^- oxidation is slowed down when soils are extremely cool and warm. High temperature also stimulates N loss to atmosphere by volatilization of N as NH_3.

Increased Precipitation

High amount of P_{cp} increases the soil water content and brings out the decrease in already dropped off O_2 supply in soil environment due to (i) low rate of O_2 diffusion into soil (at anaerobic microsites in well–aerated soil or at localized microsites of low O_2 in the center of soil aggregates); (ii) more respiratory demand by microbes than the supply inside soil aggregates; and (iii) rapid decomposition of litter. Nitrifying organisms being obligate aerobes require O_2 to form NO_2^- or NO_3^- ions, and the excess moisture in soil diminishes their activity. However, the excess moisture in soil creates conditions favorable for denitrification by inhibiting denitrifying enzyme synthesis and the electron flow to denitrifying enzyme and represses the all-nitrogen oxide reductases. The intermittent P_{cp} creates conditions favorable for nitrification. Nitrification mostly occurs during dry spell when WFPS is 40%–60%. When WFPS increased to $\geq$60%, NO_3^- thus produced are subsequently reduced to N_2 or N_2O through denitrification. So in wetting and drying, the reactive N is lost from soil as N_2O by nitrification and denitrification. During nitrification, some of the NO_3^- produced are taken by plants, and the remaining leach through percolating waters to underground waters. It may also be converted to N gas and N oxides that escape into the atmosphere. Leached NO_3^- to deeper layers is of concern for two important reasons: firstly, losses of N as NO_3^- below the root zone are not used by plants and, secondly, deterioration of drinking water quality. With sufficient P_{cp} though leaching of NO_3 is stimulated, NH_3 volatilization is ceased. Less P_{cp} or drought stimulates capillary movement of water to soil surface and enriches the soil surface with NO_3, which becomes the important source for N_2O emission.

1.3 CARBON CYCLE

Carbon cycling is constituted by atmosphere, vegetation, soil, and ocean reservoirs, which are interconnected. The CO_2 from the atmosphere enters vegetation biomass via photosynthesis (a biotic sink). A part of it is stored temporarily in vegetative tissue, and a part is lost to the atmosphere by plant respiration. Water plants, algae, cyanobacteria, and other photosynthetic bacteria that exist in agricultural system also fix CO_2. In the aboveground plants, most of the storage of C, about 75%, occurs in forests. The forest vegetation fixes CO_2, and upon senescence, eventually, C enters the soil. From the soil, CO_2 is released through soil respiration, which includes three biological processes, namely, root respiration, microbial respiration, and faunal respiration. A nonbiological process, that is, chemical oxidation, also releases CO_2, which is prominent at higher temperatures. Among the different biological respirations in soil, root respiration contributes as much as 50% of the total soil respiration. For each biological respiration, source of C supply is different. For example, for root respiration, the supply of C is made from photosynthates and its translocation to the root. For microbial respiration, C is supplied through SOM including a wide variety of organic substances ranging from freshly added substances (such as litter fall, dead roots,

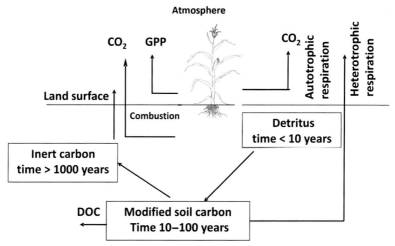

Fig. 1.3 Carbon cycle on agricultural land.

manures, and applied crop residues) at varying stages of decomposition. The decomposition of these organic substances contributes 99% of the CO_2 produced, while the contribution of soil fauna is much less. Actually in soil, bacteria, fungi, and actinomycetes act in consortium or in succession to degrade cellulose, pectin, xylan, phytin, lignocelluloses, cutin, wax, proteins, chitin, and lignin contained in the added plant and animal debris and release CO_2 ultimately. The released CO_2 is partly used by soil microbes, a part is converted to other metabolites, and a part escapes to the atmosphere. Simultaneous respiration by soil microorganisms and combustion also returns back C to the atmosphere as CO_2 gas (Fig. 1.3).

Normally under uninterrupted conditions, gain of C from above- and belowground plant biomass is lost through erosion, OM decomposition, and leaching, while simultaneously, some part is stored in the soil as C sequestration depending primarily upon the pedological factors (discussed in Chapter 5), which set the maximum limit to C storage in the soil.

1.3.1 Soil Organic Carbon and Its Pools

In terrestrial ecosystems, soil is the largest pool of actively cycling C. In soil, organic C is produced from residues of plants and animals involving a number of phenomena like decomposition, humification, accumulation, and distribution within the soil profile. Eventually, a fraction of the input organic is converted to a stable form of SOC, commonly known as humus.

The SOC from recent plant litter, to charcoal, to very old, humified compounds exists up to a depth of 1 m. Underneath 1 m depth, the CO_2 produced by anaerobic respiration, CH_4 formation, and carbonic acid reactions in soil gets solubilized in soil water and reacts with a variety of cations forming their CO_3^{2-} and is not active in circulation. The SOC is conceptually divided into four pools based on resistant to decomposition or their different turnover times (Table 1.2).

Active or labile pools are influenced by management interventions like tillage and residue and are used as indicators of changes in response to land use management. Very slowly oxidizable and passive or recalcitrant pools are more resistant and less readily impacted by agronomic practices. Resistance of SOC to decomposition is caused by chemical and physical factors. Chemically, it is the aromaticity (existence of cyclic C compounds) that determines resistance to decomposition. Higher aromaticity means more resistance of the SOC. Inert form of organic C, in charcoal, also provides resistance to decomposition. Physically, resistance to microbial decomposition is made by (i) adsorption of organic components on soil mineral surfaces and (ii) occlusion within soil structural units (micropores). In agricultural

Table 1.2 The Effect of Agronomic Practices on Soil Organic Carbon Pools

S No	Carbon Pool	Controlling Factors	Agronomic Practices
1	Active or labile pool (turnover time 1.5[a], 2.0[b] years)	Carbon form	Tillage and residue inputs and climate
2	Slowly oxidized pool (turnover time 25[a] years)	Macroaggregates	Tillage
3	Very slowly oxidized pool (turnover time 53[c] years)	Microaggregates	Have little impact on this pool
4	Passive or recalcitrant pool (turnover time 1000[a], 1429[c], 1980[b] years)	Clay mineralogy, reduction of C to elemental form by microbial decomposition	Do not influence this pool

[a]Parton et al. (1987).
[b]Jenkinson and Rayner (1977).
[c]Campbell (1978).
Modified from Eswaran, H., van den Berg, E., Reich, P., & Kimble, J. (1995). Global soil carbon resources. In R. Lal, J. Kimble, E. Levine, & B.A. Stewart (Eds.), *Soils and global change* (pp. 27–43). Boca Raton: Lewis Publishers.

soils, OM is stocked up as litter, microbial biomass (2%–5%), particulate organic matter (18%–40%), light fractions (10%–30%), intermicroaggregates (20%–25%), intramicroaggregates physically sequestered (20%–40%), and intramicroaggregates chemically sequestered (20%–40%) with estimated turnover time of 1–3, 0.1–0.4, 5–20, 1–15, 5–50, 50–1000, and 1000–3000 years, respectively (Cartar, 1996).

1.3.2 Carbon Dioxide Emission

CO_2 is produced in agricultural and forest systems through autotrophic or heterotrophic oxidation of carbon compounds (litter and organic manures) and chemical weathering of inorganic carbonate-containing primary minerals (feldspar and quartz) or secondary minerals (calcite ($CaCO_3$) or dolomite ($CaMgCO_3$)). In agriculture, CO_2 is emitted from different farm operations and use of inorganic fertilizer and pesticides in which fuel is combusted directly and indirectly. From soils, CO_2 is emitted by mineralization of SOC in which microbes use C as source of energy and release CO_2. Actually, CO_2 is increasing due to the imbalance of emission and uptake. In preindustrial period, the difference between emission and uptake was less, but afterward augmented with increasing emission by the anthropogenic activities leading to increased CO_2 in the atmosphere.

Factors Affecting Carbon Dioxide Emission

Biomass burning: Burning of surplus crop residues emits CO_2; other gases like CH_4, CO, N_2O, and SO_2; and large amount of particulate matters. According to a report of Indian Network of Climate Change Assessment, the field burning of crop residue in 2007 emitted 6.6 million ton CO_2 equivalents constituting 0.25 million ton by CH_4 and 0.007 million ton by N_2O in Indo-Gangetic Plains of India (Pathak, Agarwal, & Jain, 2012). The burning of rice straw contributes 39% of CO_2 emission. On the other hand, some scientists are of the view that emissions of CO_2 during the burning of crop residues should be considered neutral, as it is reabsorbed during the next growing season.

Fossil fuel burning: The increased atmospheric levels of CO_2 have been due to largely burning of fossil fuels by the increased population. However, a vast pool of fossil C exists beneath the earth's living layer that has remained isolated from the active C cycle for millions of years. In agriculture, CO_2 is emitted from consumption of fossil fuel by use of machinery in different operations such as pumping groundwater for irrigation, sowing of seeds, cultivation, and manufacturing and application of fertilizers and pesticides. Such emissions are more in intensive agriculture system than customary ones (Lal, 2004; Pretty, Ball, Xiaoyun, & Ravindranathan, 2002).

Tillage: Tillage emits CO_2 to the atmosphere by burning fuel, increasing rates of decomposition, or oxidizing of SOC due to increased soil aeration and direct contact of soil microbes with crop residues. In actual, impact of tillage on oxidation of SOC depends upon the availability of SOC, frequency of tillage, and soil texture. In fine-textured soils, having greater ability to retain C due to microaggregate protection, SOC gets released with the frequent tillage and redistributed in the soil profile where environmental conditions are more favorable for decomposition. As a result, the oxidation of SOC gets stimulated, and more CO_2 is emitted. In coarse-textured soils, which may accumulate practically no C, even after 100 years of high C inputs (Christensen, 1996), chances of oxidation of SOC and CO_2 emission are least for want of the sufficient availability of C. Inappropriate or excessive tillage can lead to soil degradation resulting into loss of SOC directly because of accelerated erosion and indirectly because of reduced crop yields, thereby enhancing mineralization and increased CO_2 losses.

Drainage of wetlands: Drainage of poorly drained wetlands improves soil aeration, which results in the higher rates of oxidation of SOC and emission of CO_2 to the atmosphere.

Land use changes: Land use management practices and vegetation types affect C dynamics by influencing SOC content and CO_2 flux. Forests mainly participate in natural C cycling by capturing C from the atmosphere through photosynthesis, converting photosynthate to forest biomass, and emitting C back into the atmosphere during respiration and decomposition. Forest soils can also lose a significant amount of C as a result of clearing and cultivation, forest fires, or forest management practices that reduce forest productivity. These practices reduce the C input to the soil and stimulate high rates of decomposition or soil erosion losses by increasing soil disturbances. In natural unmanaged forests, C emission is assumed to be zero as the emissions from burning, decay, and dieback are balanced by the growth of plants in these ecosystems. With deforestation, CO_2 emission is increased in both the cases, whether the system is restored for cultivation or left as such. In the former case, a lot of energy in terms of fertilizers, soil tillage, pesticides, etc. is consumed to restore the system, while in the latter case, soil gets degraded due to loss of SOC resulted from clearing of natural vegetation and reduction in the quantity of biomass returned to the soil. Under the degraded soils, CO_2 emission is more under broad leaf forest and grasslands due to higher respiration than under coniferous forest and forest stands. So changes in land use like deforestation and degradation of soils increase C losses to the atmosphere. In tropics and subtropics after clearing of forest and their conversion to farmland, approximated quarter to half of the original C in topsoil is lost as CO_2 (Lal, 1995).

Erosion: Accelerated soil erosion exacerbates CO_2 emissions. Soil erosion reduces plant productivity resulting in less biomass return to soil as above- and belowground. In addition, soil erosion reduces the SOC pool by removing C from one site and depositing it elsewhere. Nearly eroded and well-drained sloping soils with a significant water or wind soil erosion hazard often result in the higher rates of oxidation of SOC and emission of CO_2 to the atmosphere than for nearly level and well-drained soils. From erosion, 20% of the C dislocated may be released eventually into the atmosphere and 10% to the water (Lal, 1995). The emissions of CO_2 increase in tropics due to high rainfall intensity. It is estimated that through erosion-induced processes, annually emitted C into the atmosphere may be 1.14 Pg (Lal, 2001).

Soil type: Soil type affects the CO_2 emission via its SOC level. SOC level within a specific soil type varies with its composition (particle size distribution of their mineral fraction), management, biota (vegetation and organisms), topography, and frequency of various disastrous natural or human-induced events (fire, flooding, and erosion) and climate. The correlation between SOC and soil types is also explained by their zonal distribution in the world in relation to climatic factors. For example, histosols contain the highest mean SOC levels, that is, 205 t ha^{-1}, while aridosols contain the lowest, that is, 4 t ha^{-1} (Eswaran, van den Berg, & Reich, 1993). The soil type also affects the mineralization of SOM and decomposition of plant material. The net mineralization of SOM is more rapid in coarse- than fine-textured soils because of the lesser physical protection against microbial biomass attack.

Fertilizer application and use of nitrification inhibitors: Application of N fertilizers decreases CO_2 emission directly by providing N to crops and microbes and indirectly by influencing soil pH and consequently microbial activity. For example, with addition of ammonical fertilizers (NH_4NO_3), CO_2 emission is reduced as it increases acidity and reduces microbial respiration. Application of nitrification inhibitors like dicyandiamide, nitrapyrin, thiosulfate, and acetylene inhibits nitrification by acting on microbes engaged in the oxidation of OC. Calcium carbide (a slow release source of acetylene) also reduces CO_2 emission.

Cropping system and crop management: The crop management practices influence CO_2 emissions by affecting the quantity of C inputs to the soil and the quality of crop residue returned to the soil. Emission of CO_2 is increased when monocropping is replaced by multicrop rotation as there is increased SOC concentration. But some cropping systems like barley-wheat-soybean and maize-wheat-soybean have no and negative effect, respectively, for SOC accumulation and consequently CO_2 emission.

1.3.3 Effect of Climate Change on Carbon Dioxide Emission

Effect of Increased Atmospheric CO_2

With increasing atmospheric CO_2, photosynthesis process in plants is stimulated, which increases fine root biomass, the total number of roots, and root length. The increased root growth enhances the release of labile sugars, organic acids, and amino acids and stimulates microbial growth and activity. More belowground C allocation and microbial activity increase rhizosphere respiration and change the CO_2 flux, depending on the availability of nutrients (such as N). In case available N is less, increase in C released leads to immobilization of soil N, thereby limiting the N available for plants, and results in higher C:N ratio. The wider C:N ratio under elevated CO_2 favors higher fungal dominance and diversity than bacteria, which store more C than they metabolize due to higher C assimilation efficiencies. Such responses to increased CO_2 are greater below- than aboveground responses because of the activity of rhizomicrobes. Increased levels of atmospheric CO_2 can also lead to substantial increase in soil respiration. However, when C is supplied by fresh crop residues, respiration is lowered down as soil microorganisms preferentially use labile C over complex C, which redices CO_2 emission and favors C sequestration in the soil.

Effect of Increased Temperature

Increase in temperature boosts CO_2 emission, which is attributed to the increasing root activity, microbial population, and OM decomposition. Temperature being the dominant factor influences biological and non-biological (pedochemical and geochemical processes such as geothermal and volcanic soil respiration) processes and results in more CO_2 production and emission. Temperature also affects the composition of the microbial community as different microbial groups have distinct optimal temperature ranges for growth and activity. In some cases (winter soil forest/montane forest ecosystem), microbes could reduce the release of SOC owing to the loss of acclimatized microbial groups. For example, an increase in temperature in a high-latitude ecosystem decreases bacterial and fungal abundance and soil respiration, as well as a phylogenetic shift in the fungal community, suggesting that increased temperature does not always lead to enhanced C loss to the atmosphere. Effect of increased temperature on CO_2 emission also varies with the type of SOC and its sensitivity to temperature. For example, the increased temperature stimulates the use of labile C; however, recalcitrant C is relatively insensitive to temperature due to its diverse and complex structure. The increase in CO_2 emission with temperature is a matter of concern, as the possible global warming would increase CO_2 evolution from the soil that would accelerate the depletion of soil C and soil fertility. It has been pointed out by Chapman and Thurlow (1998) that rise in mean annual temperature of 5°C could potentially increase CO_2 emission by a factor of 2–4. It was also estimated that by increasing temperature of 1°C, it could reduce SOM by 10% in the regions of the world with annual mean temperature of 5°C. With the same increase in temperature (1°C), the corresponding loss would be 3% in the regions having a mean temperature of 30°C. This explains the more rapid decomposition of SOC in tropical regions than temperate. Increased temperature enhances CO_2 gas emissions also by accelerating the continual cycling of plant nutrients like C, N, phosphorus (P), potassium (K), and sulfur (S) in the soil–plant–atmosphere system.

Effect of Increased Precipitation

Precipitation affects the CO_2 emission by changing soil moisture, which is a limiting factor for soil respiration, especially at high temperatures. Dynamics of soil moisture exerts the greatest influence on soil biota and alters the microbial communities. At low soil moisture or in long periods of drier conditions due to low P_{cp}, microbial growth gets limited, and decomposition and respiration are decreased, which diminishes CO_2 emission. At high soil moisture due to more P_{cp}, CO_2 emission though increased by meeting the water demand for physiological activities of microbial communities, there is reduction due to (i) restriction of the gas diffusion rates and oxygen availability and (ii) increased plant productivity and SOC; for instance, each millimeter of P_{cp} was able to increase 48 kg ha^{-1} organic C in virgin soils in a period of 70 years (Dalal & Mayer, 1986). In fact, the net CO_2 emission depends upon the balance of these processes like microbial growth, soil respiration, and decomposition. Precipitation also affects the processes like SOM turnover, mineralization, decomposition, and controlling CO_2 emission. The combination of soil water content and temperature improves the estimates of soil respiration and CO_2 emission. More P_{cp} may also exacerbate CO_2 emissions by accelerated soil erosion and leaching (Lal, 1995). Leaching can cause C losses from any soil via DOC and dissolved inorganic carbon (DIC). In wetlands and peatlands, soil drying due to low P_{cp} may increase oxygen availability and enhance carbon cycling, thereby having a positive-feedback effect on CO_2 fluxes.

1.4 METHANE FORMATION AND EMISSION

In agriculture, the potential sources of CH_4 emission are rice fields submerged with water and have high organic C content, transportation and use of organic manure, and enteric fermentation in ruminant and grazed pastures. In the rice fields, methane is produced under the anoxic environments of submerged soils and sediments including rice paddies by the activity of methanogenic bacteria during degradation of organic C compounds. The methanogenic bacteria are anaerobes and belong to archaebacteria, which use different energy sources (acetate, methyl compounds, and format or H_2) and reduce the CO_2 to CH_4 following Eqs. (1.5a)–(1.5c), respectively. The corresponding methanogenic genera are *Methanosaeta* spp.; *Methanobacterium* spp., *Methanobrevibacter* spp., *Methanogenium* spp.; *Methanolobus* spp., and *Methanococcus* spp. Some can grow as autotrophs utilizing H_2 as energy and CO_2 as C source. The common genera are *Methanobacterium, Methanobrevibacter, Methanomicrobium, Methanogenium, Methanospirillum, Methanosarcina,* and *Methanococcus*:

$$CH_3COOH \rightarrow CO_2 + CH_4 \qquad (1.5a)$$
$$4CH_3OH \rightarrow 3CH_4 + CO_2 + 2H_2O \qquad (1.5b)$$
$$CO_2 + 4H_2 \rightarrow CH_4 + 2H_2O \qquad (1.5c)$$

The other group of microbes is methanotrophs, which is classified into three types: type I (*Methylomonas, Methylocaldum, Methylosphaera, Methylomicrobium,* and *Methylobacter*), type II (*Methylocystis* and *Methylosinus*), and type X (*Methylococcus*). Both type I and type II methanotrophs dominate in rice fields with unsaturated water content; type II methanotrophs dominated in unplanted, unfertilized soils; and type X occupies an intermediate position. Methanotrophs oxidize CH_4 by enzymatic oxidization of methane to CO_2 via methanol, formaldehyde, and formate catalyzed by the enzymes like methane monooxygenase, methanol dehydrogenase, formaldehyde dehydrogenase, and formate dehydrogenase, respectively.

During methanotrophic oxidation, energy is released; however, C is not incorporated into cellular biomass.

Rice paddies favor anaerobic activity as the soil becomes O_2-free frequently after irrigations, and the O_2 depletion favors the use of NO_3^-, Mn_4^+, Fe_3^+, SO_4^{2-}, and CO_2 as electron acceptors for respiration and decomposition of OM. The methane production is initiated at an Eh of -150 mV and becomes more at -200 mV utilizing the H_2, CO_2, format, acetate, ethylated amines, and methanol. Anaerobic fermentation produces main gases like H_2, H_2S, N_2, CH_4, and CO_2. The produced CH_4 is transferred to atmosphere by the mechanisms of (i) diffusion; (ii) ebullition; and (iii) plant-mediated transport via the aerenchyma of vascular plants (Fig. 1.4). These three mechanisms regulate the spatial and temporal variations in CH_4 production. In the first process, diffusion takes place because of the formation of a CH_4 concentration gradient from deeper soil layers, where the production of CH_4 is large, to the atmosphere. It is a slow process compared with the other two transport mechanisms, but is a biogeochemically important as it extends the contact between CH_4 and methanotrophic bacteria in the upper aerobic layer and promotes CH_4 oxidation. In the second process, ebullition (formation of gas bubbles in flooded soil and emigration to surface) takes place when CH_4 production is large. As this process is fast, CH_4 oxidation is absent or negligible. In the third process, plant-mediated transport takes place through an internal

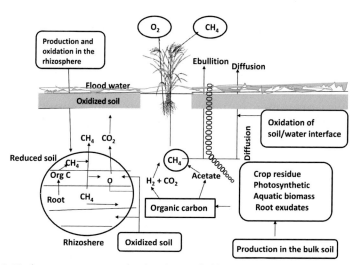

Fig. 1.4 Methane transport mechanism in rice fields.

system of continuous air spaces named aerenchyma, a structure developed by vascular plants to adapt to flooded environments for the transport of O_2, necessary for root respiration and cell division in submerged organs. But this structure is also used for CH_4 transportation from rhizosphere to atmosphere by passing the aerobic CH_4 oxidizing layers. This process involves two major mechanisms, molecular diffusion and bulk flow. In the molecular diffusion, gradient of CH_4 concentration formed inside the aerenchyma conduits acts as the driving force for CH_4 diffusion from the peat root zone to the aerial parts of the plant. The other transport mechanism is bulk transportation, in which driving force is a pressure gradient generated by differences in temperature-induced water vapor pressure between the internal air spaces in plants and the surrounding atmosphere. By this mechanism, transport of CH_4 occurs along the plant, also through the aerenchyma structure, from the leaves to the rhizome and back to the atmosphere through old leaves or horizontal rhizomes connected to other shoots. At global level, irrigated, rain-fed, and deep-watered rice contributes 70%–80%, 15%, and 10% of CH_4 emission, respectively, from rice agriculture (Wassmann et al., 2000).

Methane is also produced as a direct result of anaerobic decomposition processes during storage, handling, and field application of manures and is influenced by the composition of the manure (amount of degradable organic compounds) and storage conditions. Temperatures of air and manure, moisture, pH, and manure storage or residency time affect the amount of CH_4 produced because they influence the growth of the bacteria responsible for CH_4 formation. The much larger production of CH_4 during manure storage (especially in liquid form) is attributed to the continued fermentation of the manure compared with major aerobic decomposition of fecal excretions in the field. Methane emissions appear to be greater under wetter and colder climatic conditions (Chadwick, Pain, & Brookman, 2000).

Enteric fermentation in ruminant is another major source of CH_4 emission. Enteric CH_4 produced principally from microbial fermentation of hydrolyzed dietary carbohydrates such as cellulose, hemicelluloses, pectin, and starch and emitted by an animal is influenced by dietary factors (type of carbohydrate in the diet; level of feed intake; level of production, e.g., annual milk production in dairy; digest passage rate; the presence of ionophores; and degree of saturation of lipids in the diet), environmental conditions (temperature), and genetic factors such as efficiency of feed conversion (Kebreab, Clark, Wagner-Riddle, & France, 2006). Ruminant animals (e.g., cattle, buffalo, sheep, and goat) have the highest CH_4 emissions among all animal types because of their unique digestive system and large intestine composed of a rumen or large "forestomach."

Small amounts of CH_4 can also be produced in grazed pastures, from animal excreta deposited during grazing, with a majority produced in dung patches (Kebreab et al., 2006). Methane emissions from dung vary depending upon its type, time since deposition, and local climatic conditions at the time of deposition, while soil type appears to have only a limited effect (Saggar et al., 2004).

1.4.1 Factors Affecting Methane Emission

Any factor that controls redox potential in soil, competition of sulfate-reducing bacteria and methanogens, availability of substrate for methanogens, and oxidation of CH_4 affects the CH_4 gas emission. The factors affecting are the following:

Chemical N fertilizer. Nitrogen fertilizers affect variably on CH_4 emissions. Application of both urea and ammonium sulfate reduces CH_4 emissions, but reduction is more with ammonium sulfate due to the competition of sulfate-reducing bacteria with methanogens. Ammonium sulfate reduces CH_4 emissions more compared with potassium nitrate as the nitrate fertilizers increase redox potential by adding the NO_3^-. This implies that to reduce CH_4 emission, there is a need to maintain redox potential (from -150 to -200 mV) favorable to sulfate-reducing bacteria. Manures applied in combination with urea minimize CH_4 emission in upland rice. In water-flooded rice soil, the type and the mode of application of a fertilizer affect CH_4 emissions. Organic manures greatly promote CH_4 emission as compared with other fertilizers, while mineral fertilizer releases the least. The CH_4 emission is lower with a mixture of prilled urea and nimin (a nitrification inhibitor) than with no fertilizer and prilled urea alone (Rath et al., 1999).

Flood water regimes and depth of water table: In flooded soils, methanogenic activity is high as the redox potential is favorable for methanogens. In natural wetlands, the type of vegetation and the depth of the water table are the factors that have an important effect on CH_4 emissions; however, the effect of the latter is more and positively correlated with CH_4 emissions. The plants that emit more CH_4 are *Carex lasiocarpa*, *C. meyeriana*, and *Deyeuxia angustifolia*. Methane emission from various rice ecosystems differs following the order of irrigated continuous flooding > rain-fed > irrigated intermittent flooding > deep water (Bhatia, Pathak, & Aggarwal, 2004; Wassmann et al., 2000).

Type and activity of vegetation: The three functions of vegetation deter-mining CH_4 emission are (i) root exudation; (ii) methane transport to atmosphere; and (iii) CH_4 oxidation in soil. Plant root system continuously releases a wide range of labile C compounds, that is, mucilage, ectoenzymes, organic acids, sugars, phenolics, and amino acids. These root exudates are used easily by methanogenic microorganisms and have substantial positive effect on CH_4 production in the soil. At the same time, some plants like rice may transport CH_4 and O_2 from the air to the root zone, via aeren-chyma. A high diffusion capacity of aerenchyma entails higher efflux of CH_4 plus higher oxidation power of the root in the rhizosphere. Addition of C compounds by vascular plants in peat soils with a deep water table, where O_2 supply is scarce, is of no importance as that had already more than 90% OC.

Organic matter: The concentration and type of OM and the concentration of O_2 are considered the determining factors for CH_4 production. The presence of some compounds can be unfavorable for methanogenesis. For example, composted organic sources (Azolla compost, blue-green algae compost, and farmyard manure) had less effect on the production of CH_4. Application of poultry manure results in low emission of CH_4 as its high sulfur content inhibits methanogenic microorganisms because of the competition with sulfate reducers for substrates such as hydrogen (H_2) and acetate and the toxicity of sulfide formed during anaerobiosis. However, the additions of fresh organic materials (rice straw and other cellulosic mate-rials) to a rice soil enhance CH_4 production and emission as they serve as C substrates for soil microorganisms.

Tillage: Tillage practices may reduce CH_4 emission by affecting soil's capacity to oxidize atmospheric CH_4 due to direct or indirect impacts of mechanical disturbance on soil N availability and soil physical properties like porosity and soil structure (Ussiri, Lal, & Jarecki, 2009). Impact of tillage is almost lacking on CH_4 oxidation (Suwanwaree & Robertson, 2005). But under no-tillage with stubble retention in irrigated cropping systems, CH_4 emission may be more because higher WFPS coupled with the presence of easily decomposable crop stubble in no-tillage system may create ideal conditions for greater CH_4 production than oxidation in soil. However, with no stubble retention, there may be reduction in soil CH_4 emission because of improved soil macroporosity and pore continuity, which enhance CH_4 diffusivity and uptake by methanotrophs (Ussiri et al., 2009).

Pesticide application: Pesticides reduce CH_4 production by inhibiting methanogenesis. A significant inhibition in CH_4 fluxes, both plant mediated and ebullition, in rice flooded field by the herbicide butachlor applied at 1 kg ai L^{-1} and by Tara-909 (dimethoate 300 g kg^{-1}) in vitro is documented (Mohanty, Nayak, Babu, & Adhya, 2004). Other pesticides like organochlorine insecticide dichlorodiphenyltrichloroethane (DDT) and carbofuran, a carbamate insecticide, are also known to inhibit CH_4 production in rice.

Effect of Climate Change on Methane Emission
Effect of Increased CO_2
Increased CO_2 levels lead to an increased CH_4 emission due to four reasons: (i) increases methanogenesis; (ii) decreases CH_4 uptake substantially (up to 30%) by soil microorganisms; (iii) decreases CH_4 oxidation because of decreased soil redox potential; and (iv) increases C substrate. Increased CO_2 levels also affect the CH_4 emission in the soil through increased soil moisture owing to reduced transpiration, which lead to more anoxic conditions, thereby increasing methanogenesis and reducing methanotrophy. In case CH_4 emission decreases with increased atmospheric CO_2, it is due to abundance and community structure of the methanotrophs, which transform the CH_4 to CO_2. The positive effects of CO_2 on plant growth and soil C input also stimulate CH_4 emission. From a metaanalysis, using 152 observations from 49 published studies, it has been concluded by Van Groenigen et al. (2011) that increased CO_2 (ranging from 463 to 780 ppm) stimulates emissions of CH_4 by 13.2% from rice paddies and natural wetlands.

Effect of Increased Temperature
Under warmer conditions, when other factors are not limiting, activities of both methanogens and methanotrophs are increased, and CH_4 emission is stimulated. For instance, a global rise in temperature of 3.4°C has been predicted to increase CH_4 emissions from wetlands by 78% (Shindell, Walter, & Faluvegi, 2004). Effect of warming is influenced by soil aeration status. Warming may lead to a substantial increase in net CH_4 emissions from anaerobic permafrosts and wetlands at high latitudes because of partial inhibition of microbial respiration, emerged from inactivation of biological oxidation systems due to decreased soil redox potential. However, in aerobic surface soils (oxygen and atmospheric CH_4), CH_4 emission from soil is decreased due to increased CH_4 oxidation rates and net CH_4 uptake by soil

microorganisms with increased gas diffusion rates and microbial access. Increasing temperature generally offsets the stimulatory effect of increased CO_2 levels on CH_4 flux.

Effect of Increased Precipitation

Increased P_{cp} enhances methanogenesis and net CH_4 emissions by increasing net primary production, increasing soil water contents, and decreasing water table depths, whereas reduced P_{cp} and drying of soils promote oxygen availability and CH_4 oxidation, which reduce net CH_4 emissions. For example, some studies showed decline of type II methanotrophs in response to increasing P_{cp} and temperature. However, the amount of CH_4 formed in permafrost tundra environments observed the opposite effect of increased temperature on these microorganisms (Knoblauch, Zimmermann, Blumenberg, Michaelis, & Pfeiffer, 2008).

Though the impact of individual climate variable on CH_4 emission is known, there is a need to understand net effect under the changed P_{cp} simultaneously with other climate parameters like CO_2 and temperature. In one of such studies, it has been reported that though the stimulatory effect of temperature and CO_2 was indirect and mediated through their impact on soil moisture, but the effect of increased CO_2 levels on CH_4 flux was offset by increasing temperature because increased temperature enhances CH_4 uptake, whereas increased CO_2 levels decrease it.

1.5 SIMULTANEOUS EMISSION OF CARBON AND NITROGEN GASES

In real agricultural systems, changes in ecological factors and management interventions may emit C and N gases simultaneously. For example, when soil gets rewetted partially after drying, microbial activity is improved, and emissions of N_2O and CO_2 get peaked. Such behavior is named as the "pulse" or "birch" effect (Jarvis et al., 2007). The N_2O pulses are mainly driven by denitrification. When C and N are added simultaneously in a soil, N gets immobilized by the microbial biomass, and there is increase in C availability, which reduces overall N_2O emissions. All these biochemical processes can also occur simultaneously in the soil due to variation in levels of soil parameters such as WFPS, NH_4^+, and NO_3^- or sequestered C at different microsites. In soil, the potential hot spots for N_2O production are microaggregates as nitrifier and denitrifier communities are larger and fluctuate more at this point.

1.6 MEASUREMENT AND ESTIMATION OF GREENHOUSE GASES

The GHG emission is either measured in the field or estimated by using developed empirical relations and complex models.

1.6.1 Field Measurements

Under field conditions, NO_2, CO_2, and CH_4 fluxes are generally measured using the static and closed chambers and micrometeorologic methods.

Static and closed chamber methods: These are open-bottom cylinders placed on soil surface for a sufficient time to accumulate the emitted gas from the soil within the enclosed head space. The gas sampling is performed periodically at 0, 30, and 60 min from head space with syringe, and the linear portion of the accumulation curve corresponds the gas flux rate. The analyses of soil air samples for CO_2, N_2O, and CH_4 are conducted using a gas chromatograph (GC). The GC is equipped with a thermal conductivity detector for CO_2 analysis, 63 Ni electron capture detector for N_2O, and a flame ionization detector for CH_4 analysis. Standard CO_2, CH_4, and N_2O samples are used for GC calibration. Diurnal fluxes (q) of gases ($mg\ m^{-2}\ day^{-1}$) are calculated by Eq. (1.6):

$$q = ((\Delta CO_2 - C \text{ or } \Delta N_2O - C \text{ or } \Delta CH_4 - C)/\Delta T)(V/A)\,k \quad (1.6)$$

where ΔCO_2—C or ΔN_2O—C or ΔCH_4—C is the linear change in gaseous concentration ($mg\ m^{-3}$) inside the chamber, V is the chamber volume (m^3), A is the surface area of the chamber (m^2), T is time (days), and k is the time conversion factor ($1440\ min\ day^{-1}$). The chamber gas concentrations in parts per million (ppm) determined by GC analysis are converted to mass per volume by assuming ideal gas relation. During each gas sampling, soil samples from 0 to 5 cm are collected to determine the soil moisture content. The advantages of static flux chambers are less costly, easy to use, and useful for assessing short-term changes in gas flux from specific environment. But due to their limitations such as (i) covering only a small surface area; (ii) having to be left in place too long to capture the effects of the soil microclimate and gas fluxes; and (iii) having more chances of missing an important flux event, sophisticated and automated chambers are preferred over these (Meyer et al., 2001). Though maintenance of automated chambers is relatively expensive yet these have the advantage of continuous measurements with less labor and without missing important events.

For measurement of CO_2 emission, infrared analyzer and soil respirator methods are also used.

Infrared analyzer method: In this method, the CO_2 emission is measured with an environmental gas monitor (EGM) chamber in a closed dynamic chamber attached to a data logger. The chamber is placed at the soil surface for 2 min for each plot until CO_2 emission measurement is recorded in the data logger. The increase in CO_2 concentration in the enclosed chamber over a specified time estimates CO_2 emission. Soil CO_2 flux is determined when the chamber CO_2 concentration equals the ambient atmospheric concentration.

Soil respirator method: In this method, a closed chamber of known volume, called soil respiration chamber, is placed on the soil, and rate of increase in the CO_2 concentration inside the chamber is monitored by EGM. With this system, the air is continuously sampled in a closed circuit through the EGM, and the soil respiration rate is calculated and recorded. A quadratic equation is fitted to the relationship between the increasing CO_2 concentration and elapsed time. The flux of CO_2 per unit area and per unit time is measured from Eq. (1.6).

Micrometeorological Method

Fluxes of GHGs from the agricultural fields are measured by Eddy covariance (EC) flux tower, which is a micrometeorologic technique. It is based on measurements of vertical turbulent flux of water, CO_2, heat, CH_4, ozone, and volatile organic components in the atmospheric boundary layer in air over ecological unit. This approach has the improvement over chambers as it (i) measures the fluxes continuously and for a larger area (several million hectares) without disturbance; (ii) measures canopy interaction with gas fluxes; and (iii) integrates spatial and temporal heterogeneity in fluxes (Hendriks, Dolman, van der Molen, & van Huissteden, 2008). Movement and gas content of air above and within the plant canopy are detected by sensors mounted on the tower. Three-axis (3-D) sonic anemometer and a fast-response infrared gas analyzer are used to measure CO_2 and H_2O concentrations at a point over the canopy. With these sensors, annual fluxes are quantified for the entire landscape unit. The major limitations of micrometeorological method are high operational cost and technical limitations associated with weather conditions (related to boundary layer conditions, air turbulence, and period of P_{cp}). Thus, to have important information about temporal fluxes, especially across environmental events that create high fluxes, the use of both automated chambers and micrometeorologic approaches is to be made. For more detailed information on measurement of greenhouse gases, refer Dalal, Wang, Robertson, and Parton (2003) and Pathak, Upadhyay, Muralidhar, Bhattacharyya, and Venkateswarlu (2013).

1.6.2 Estimation From Empirical Relationships and Models
Nitrous oxide

Generally, GHG emission data are articulated as emission factor (EF) and cumulative emission. For N_2O, EF is the difference in N_2O emitted from fertilized and unfertilized treatments per unit of fertilizer applied to soil during the measurement period (Eq. 1.7):

$$EF = \left(\frac{kg\,N_2O\,em\,(\text{fertilized}) - kg\,N_2O\,em\,(\text{unfertilized})}{kg\,N\,\text{fertilizer applied}} \right) \quad (1.7)$$

The EFs vary with type of fertilizer and irrigation methodology (Eduardo et al., 2013). From a number of experiments in Mediterranean climate, the median values of EF were 0.48, 0.59, 0.86, and 1.18 for organic solids, synthetic, organic liquid, and organic + synthetic fertilizers, respectively. For rain-fed, high water, and drip irrigation, the values were 0.08, 1.01, and 0.66, respectively.

For CO_2, EF is expressed by Eq. (1.8):

$$EF = \frac{(C\,\text{sample} \times Q\,\text{chamber} \times t)}{M\,\text{burned}} \quad (1.8)$$

where EF is in $mg\,kg^{-1}$, C sample is the concentration of the GHGs in the sample $(mg\,m^{-3})$, Q chamber is the flow rate of dilution air into the chamber $(m^3\,min^{-1})$, t is the burn sampling time in minutes, and M burned is the mass (kg) of biomass burned.

Cumulative emission is the sum of N_2O fluxes over a given time period (may be 1 year). Since rates of denitrification vary greatly in space and time, this makes it very difficult to measure in field conditions. However, Bouwman (1996) developed a regression equation taking long-term measurement data sets of N sources from a variety of mineral and organic N fertilizers for estimation of total emissions of N_2O and N fertilizer applied as Eq. (1.9):

$$\text{Annual direct field } N_2O - N\,\text{loss} = 1 + (0.0125 \times N - \text{application}) \quad (1.9)$$

The value of 1 in the equation represents background emission evolved $(kg\,N_2O{-}N\,ha^{-1})$, while 0.0125 factor accounts for the contribution from applied fertilizer $(kg\,N\,ha^{-1}year^{-1})$. Similarly, annual denitrification in

Danish agricultural soils, based on a combination of average results from the literature, several years of experience, and a portion of common sense, was estimated by Vinther and Hansen (2004) employing SimDen model. In Sim-Den model, denitrification is calculated as Eq. (1.10). The SimDen model was modified as SimDen-clay by incorporating the N_2O emission factor (Eq. 1.11) as a function of WFPS and hydraulic conductivity of the soil assuming that the effects of these parameters are integrated in clay content of the soil and have a significant effect on emission factor:

$$\text{Denitrification} = (N_2O \text{ emission}) \times \left(\frac{N_2}{N_2O}\right) \qquad (1.10)$$

$$\text{Denitrification} = \left(\begin{array}{c}\text{Background } N_2O \text{ emission} + \\ (N \text{ input} \times N_2O \text{ emission factor})\end{array}\right) \times \left(\frac{N_2}{N_2O}\right) \quad (1.11)$$

Based on such studies, IPCC (2006) proposed a linear relationship between N_2O fluxes and rates of fertilizer application assuming that total input of N to soil and its subsequent availability is the robust predictor of N_2O fluxes. Contrary to IPCC (2006) proposal of linear relation between N_2O emission and fertilizer rate, a number of studies suggest that this relationship is not linear and EF varies with fertilizer rate. The EF is lowest at low N fertilizer rates (0–75 kg N ha^{-1}) and highest at high rates (200–225 kg N ha^{-1}). Little effect of small fertilizer additions and disproportionate of levels greater than crop need on N_2O emissions are documented in a recent multiple N-rate experiment (Ruan, 2014). Likewise, different relations have been recognized in different experiments, for example, out of 26 total studies, the response was linear in 6, exponential in 18, and hyperbolic in 2 (Kim, Hernandez-Ramirez, & Giltrap, 2013). A faster than linear N_2O emission increase (Eq. 1.12) for US Midwestern maize crops with a model based on log-transformed values was recognized by Hoben, Gehl, Millar, Grace, and Robertson (2011):

$$N_2O_{\text{emission}} = \frac{[6.7(e^{0.0067}N - 1)]}{N} \qquad (1.12)$$

Later on, using nontransformed emissions and from multiple N input data, the following relations (Eqs. 1.13–1.17) have been developed:

$$\text{Millar et al. (2013) } N_2O_{\text{emission}} = (4.00 + 0.026\,N)N \qquad (1.13)$$

$$\text{Hoben et al. (2011) } N_2O_{\text{emission}} = (4.36 + 0.025\,N)N \qquad (1.14)$$

$$\text{Shcherbak, Millar, and Robertson (2014) } N_2O_{\text{emission}} = (6.49 + 0.0187\,N)N$$

$$(1.15)$$

$$\text{Shcherbak et al. (2014)} \quad N_2O_{\text{emission}} = (6.58 + 0.0181N)N \atop \text{(Excluding N-fixing crops)} \tag{1.16}$$

where N is input in kilogram (kg) of N ha^{-1} and N_2O emission is gram (g) of N_2O—N ha^{-1}. On an average, the emissions are best characterized by the relation (Eq. 1.17):

$$N_2O_{\text{emission}} = (6.58 + 0.0181\,N)N \tag{1.17}$$

Similarly, Pathak and Wassmann (2007) developed relations for CO_2 (Eq. 1.18), CH_4 (Eqs. 1.19, 1.20), and N_2O (Eq. 1.21) emissions incorporating soil factors (SOC, bulk density, and soil depth), crop duration and rate of decomposition, root input + manure input, temperature correction factor, and a technology-specific index (Table 1.3) as follows:

CO_2 emission

$$
\begin{aligned}
\left(CO_{2\_}em\_ac,\, kg\,C\,ha^{-1}\right) = {}& (SOC,\, per\,cent) \times 1000 \\
& \times \left(bulk\,density,\, g\,m^{-3}\right) \times (soil\,depth,\, cm) \\
& \times 0.000085 \times (crop\,duration,\, days) \\
& \times 2** \frac{\left(Temp,\,^{\circ}C - 25\right)}{10} \times Tech\_SOC\_CO_2
\end{aligned}
\tag{1.18}
$$

$$
\begin{aligned}
\left(CH_{4\_}em\_ind,\, kg\,C\,ha^{-1}d^{-1}\right) = {}& SOC \times 1000 \times bulk\,density \times soil\,depth \\
& \times 0.000085 \times crop\,duration \times 0.27 \times 0.55
\end{aligned}
\tag{1.19}
$$

Table 1.3 Technology-Specific Index for Carbon Dioxide Emission by Different Farm Inputs and Operations

Input	CO_2 Emission (kg C kg^{-1}) Input	Operation	CO_2 Emission (kg C kg^{-1}) Input
Nitrogen	1.3[a]	Mouldboard plowing	15.2[b]
Phosphorus	0.2	Field cultivation	4.0
Potassium	0.2	Sowing	3.2
Herbicide	6.3	Fertilizer spreading	7.6
Insecticide	5.1	Spray pesticide	1.4
Fungicide	3.9	Irrigation per 100 mm	5.0
		Harvesting	10

[a]Schlesinger (1999).
[b]Singh et al. (1999) and Lal (2004).

$$\left(CH_4\_em\_ac, kg\,C\,ha^{-1}d^{-1}\right) = [CH_4\_em\_ind \times Tech\_CH_4$$
$$+(root\,input + manure\,input \times 0.5)$$
$$\times 0.27 \times 0.55 \times 0.4]$$
$$\times 2** \frac{(Temp, ^\circ C - 25)}{10} \qquad (1.20)$$

$$N_2O\_em\_ac = \left[\frac{(CO_2\_em\_ac + CH_4\_em\_ac)}{10} + (FertilizerN)\,x\,0.024\,x\,Tech\_N_2O\right]$$
$$(1.21)$$

where Tech_CH$_4$ is a technology-dependent factor for CH$_4$ emission, root input and manure input correspond to the respective organic input (kg), 0.5 represents the fraction of manure mineralized during the growing season (assuming that 50% of the manure will be decomposed during the fallow period), 0.27 is the ratio of molecular weights of methane and carbohydrate, 0.55 is the initial fraction of emitted methane, 0.4 is the C content of the root and manure inputs, and 2**(Temp $^\circ$C $-$ 25)/10 is the temperature correction factor, where Temp is the seasonal average temperature ($^\circ$C).

These derived empirical relationships (Eqs. 1.9–1.17) explain well the N$_2$O emission with the amount of fertilizer N applied, but are independent of fertilizer type, weather (P_{cp} and temperature), period covered by the measurement, the presence and type of crop, crop residue, tillage, mode and timing of fertilizer application, soil properties, etc. To have the integrated effect of soil, crop, climate, and temporal variation, process-oriented multicompartment models like DAYCENT (Parton, Hartman, Ojima, & Schimel, 1998), DNDC (Li, Frokling, & Frokling, 1992), NLOOS (Riley & Matson, 1989), and Expert N (Baldioli et al., 1994; Engel & Priesack, 1993) for N and CENTURY-NGAS (Parton et al., 1996; Parton, Schimel, Cole, & Ojima, 1994), NASA-CASA (Potter, Riley, & Klooster, 1997), and RothC for C have been developed. In these models, ecological drivers (climate, soil physical properties, vegetation, and anthropogenic activity), soil environment variables (soil temperature, moisture, pH, Eh, and substrate concentration), and gas (NO, N$_2$O, CH$_4$, and NH$_3^-$)-related biogeochemical reactions are integrated into one framework. Among these models, DAYCENT, DNDC, RothC, and CENTURY are commonly used. The characteristics of these models are summarized in Table 1.4. Each of the models has its own strategy or philosophy to use number of input parameter and empirical equations for capturing basic patterns of gas fluxes to include more mechanisms for better tracking processes affecting

Table 1.4 Characteristics of the Models

	DAYCENT, Parton et al. (1998)	DNDC, Li et al. (1992)	Roth-C	CENTURY, Parton et al. (1987)
Atmosphere	Daily rainfall, maximum and minimum temperatures	Daily rainfall, maximum and minimum temperatures, initial concentrations of the substrates	Monthly rainfall, maximum and minimum temperatures, potential ET	Monthly rainfall, T_{max}, T_{min}
Plant	Plant growth model simulates net plant productivity as a function of genetic potential, phenology, nutrient availability, water/temp stress, and solar radiation. Allocate NPP to plant components (e.g., roots vs shoots)	The plant growth submodel simulates daily water and N uptake by vegetation, root respiration, and plant growth and partitioning of biomass into grain, stalk, and roots	NO plant model DPM/RPM ratio = decomposable plant material/resistant plant material	No plant production model
Soil	Carbon model simulates SOM in the top 20 cm soil layer as a sum of dead plant matter. Simulate decomposition of biomass, and the decomposed products enter the SOM pools (active, slow, and passive) and flow between the pools depending on lignin content and C:N ratio, size of the pools, temperature/water factors, and clay content. Simulate nutrient	The soil climate submodel simulates soil temperature and moisture profiles based on soil physical properties, daily weather, and plant water use. Simulate SOC residing in four major pools: plant residue (i.e. litter), microbial biomass, active humus, and passive humus. Each pool consists of 2 or 3 subpools with different specific decomposition rates	Clay content (%), monthly soil cover (whether the soil is bare or vegetated). Monthly input of plant residues (t C ha^{-1}). Monthly input of FYM (t C ha^{-1}) if any	Soil texture (% sand, silt, and clay). Bulk density (0–20 cm surface soil). Initial SOM (g C m^{-2}), initial pools, active surface, active soil, slow soil and active soil C:N ratio, initial SOM-N, initial SOM-C, soil texture regulates the C transfer from one compartment to the other

Continued

Table 1.4 Characteristics of the Models—cont'd

	DAYCENT, Parton et al. (1998)	DNDC, Li et al. (1992)	Roth-C	CENTURY, Parton et al. (1987)
	mineralization as functions of substrate availability, lignin content, C:N ratio, water/temperature stress, and tillage intensity			Simulate long-term SOM dynamics C, N, P, and S dynamics
Output	Denitrification and nitrification Daily denitrification rates are calculated for each soil layer based on soil NO_3 concentration distributed throughout the soil profile, heterotrophic respiration (i.e., available labile C), soil water content, texture, and temperature, while nitrification rates are calculated based on soil NH_4 concentration, water content, texture, and denitrification temperature in the top 15 cm layer	The decomposition submodel simulates daily decomposition, nitrification, ammonia volatilization, and CO_2 production by soil microbes The submodel calculates turnover rates of soil organic matter at a daily time step The nitrification model tracks growth of nitrifiers and turnover of ammonium to nitrate The denitrification submodel operates at an hourly time step to simulate denitrification and the production of nitric oxide (NO), N_2O, and dinitrogen (N_2) CH_4 production and oxidation are simulated under anaerobic conditions	Simulates changes in SOC stock on a monthly basis Calculate the inert organic matter from SOC (Falloon et al., 1998) and convert SOC to t C ha^{-1}, from sand, clay, and organic matter content (Dean et al., 1989)	

gas production/consumption. Each model has its own strength and limitation. For example, the strength of DNDC is as follows: it works well for simulating emission of NO, N_2O, and CH_4 in upland and waterlogged conditions. Limitations are as follows: (i) Its capacity of predictions is limited to the plant-soil system within the field scale of cropland; (ii) at regional scale, a number heterogeneity of soil properties, that is, texture, SOC content, and pH cause uncertainty; and (iii) the current of DNDC does not track the fate of DOC in underground water or streams and is not capable to simulate aquatic biogeochemistry. However, Li et al. (2004) gave the most sensitive factor (MSF) approach to reduce the magnitude of the uncertainty for rice crop by modifying growth rates of vegetative and reproductive stages in terms of accumulative thermal degree days to adjust the rice-growing season. Similarly, the strength of RothC is as follows: it (i) simulates the turnover of C in nonwaterlogged soil as affected by soil type, temperature, moisture, and plant cover.

Similarly, the strength of Roth C is as follows: it (i) simulates the turnover of C in nonwaterlogged soil as affected by soil type, temperature, moisture, and plant cover; (ii) performs two types of simulations: "direct" that uses the known input of OC to the soil to calculate the SOC and "inverse" that evaluates the input of OC required to maintain the stock of SOC; and (iii) requires limited and easily available input data. Limitations are the following: (i) simulate changes in SOC stock on a monthly basis and (ii) used for upland crops only.

1.7 PAST AND FUTURE TRENDS OF GREENHOUSE GASES EMISSION

Nitrous Oxide

Atmospheric N_2O concentration at global level was 270 ppb (parts per billion) in 1900, which increased by 17% from 1990 to 2005. It is projected that due to increased N fertilizer use and increased animal manure production, it may increase by 35%–60% up to 2030 (FAO, 2003). In India, increase in the N_2O emissions observed from 1980 to 2007 was 176% (Bhatia, Jain, & Pathak, 2013).

Carbon Dioxide

Atmospheric CO_2 concentration of 270 ppm in 1750 (preindustrial period) has reached 400 ppm in 2014 with intermittent increase of 1.5 ppm year^{-1}

in the 1980s and 1990s and 2.0 ppm year^{-1} during the last decade (2006–15). The annual CO_2 increase during the past year (2015) was 3 ppm for only the second time since 1979 (Butler & Stephan, 2016). For future concentrations, the SERES scenario has led to divergent concentration trajectories. Among the six scenarios considered, the projected range of CO_2 concentration at the end of the century is 550–970 ppm from ISAM model (Jain, 2000) or 540–960 ppm from Bern CC model. NOAA (2008) projected that CO_2 is expected to reach 450–550 ppm in 2050 and 700 ppm by the end of the 21st century due to the uncontrolled human activities such as burning of fossil fuels, increased use of refrigerants, and enhanced agricultural activities. Watson, Rodhe, Oeschger, and Siegenthaler (1990) projected that atmospheric CO_2 concentration of 350 ppm in 1990 will reach between 600 and 1000 ppm by 2100. Agricultural scientists are of the view that future CO_2 emission due to nonagricultural activities would increase, but from agriculture sector, emission is likely to decline or remain at low level due to adoption of adaptation technologies and mitigation strategies. India's contribution of CO_2 to the world total is only 4.6% compared with the United States's contribution of 20.9% followed by 17.3% of China (GOI, 2015).

Methane

At global scale, agricultural CH_4 methane emission increased by 17% from 1990 to 2005 (US-EPA, 2006). However, growth rate declined from 1983 to 1999, remain constant from 1999 to 2006, and increased since 2007 due to warm temperatures in the Arctic and increased P_{cp} in the tropics in 2007 (Butler & Stephan, 2016). From 2014 to 2015, global CH_4 emission increased faster (11.5 ppb year^{-1}) than it had from 2007 to 2013 (5.7 ± 1.2 ppb year^{-1}). In 2015, it has increased to 1850 ppb. By 2030, CH_4 is expected to increase 60% with the increased livestock (FAO, 2003). It is forecasted that CH_4 emission from enteric fermentation plus manure management and rice crop will increase by 21% and 16%, respectively, in 2005–20 (US-EPA, 2006). In India, from 1980 to 2007, there was almost no increase in CH_4 (Bhatia et al., 2013).

1.8 WARMING EFFECT OF GREENHOUSE GASES

As the GHGs absorb outgoing long-wave terrestrial radiation from the earth's surface, it leads to rise in air temperature and causes global warming. Each GHG has its own global warming potential (GWP), which is an index used to compare the effectiveness of the GHG in trapping heat in the atmosphere relative to a standard gas, CO_2 by convention. On the global climate system, warming effect of GHGs differs due to their different magnitudes, lifetimes in the atmosphere, and radiative properties. For example, among the GHGs, N_2O is a powerful gas. In terms of radiative forcing, N_2O is 310 times and CH_4 is 21 times powerful than CO_2 (based on a 100-year time horizon). The N_2O also has a long lifetime, that is, 120 years in the atmosphere in comparison with 60 years of CH_4 and 10 years of CO_2 (Shine, Derwent, Wuebbles, & Morcrette, 1990), and is very steady. The contribution of N_2O, CO_2, and CH_4 to global warming is 5%, 60%, and 15%, respectively (Watson, Zinyowera, & Moss, 1996). GWP is calculated using Eq. (1.22), given by IPCC (2006):

$$
\begin{aligned}
\text{GWP}\left(\text{kg}\,CO_2\text{equivalent}\,\text{ha}^{-1}\right) = {} & CH_4\_em\_ac \times \left(\frac{\text{molecular weight of methane}}{\text{molecular weight of carbon}}\right) \\
& \times \text{radiative forcing of methane compared } to \text{ carbon} \\
& + N_2O\_em\_ac \times \left(\frac{\text{molecular weight of nitrous oxide}}{\text{molecular weight of carbon}}\right) \\
& \times \text{radiative forcing of nitrous oxide compared } to \text{ carbon} \\
& + (CO_2\_em\_ac + CO_2\_em\_op + CO_2\_em\_in) \\
& \times \left(\frac{\text{molecular weight of carbon dioxide}}{\text{molecular weight of carbon}}\right) \\
& \times \text{radiative forcing compared } to \text{ carbon} \quad (1.22)
\end{aligned}
$$

EXERCISES

1. Thrash out the individual and interactive effects of changed climate parameter, that is, increased CO_2, temperature, and precipitation on the emissions of nitrous oxide, carbon dioxide, and methane gases.
2. Do greenhouse gases (GHG) change climate or climate changes GHG emission? Which is stronger and why? Discuss.
3. Why (i) denitrification is more in rhizosphere than nonrhizosphere and in fine- than coarse-textured soil and (ii) leaching of N is more with application of organic fertilizers than inorganic?
4. Denitrification is an undesirable reaction for agricultural productivity, but it is of major ecological importance, discuss.
5. Assuming that Eq. (1.9) in this chapter holds true for estimating N_2O emission in rice–wheat cropping system, calculate the percentage of N fertilizer applied (120 kg ha^{-1} to rice and 120 kg ha^{-1} to wheat) that get emitted as N_2O annually.

Answer (1.67%)

6. Will proportion of N_2O emission to N level is same if multilevels of N fertilizers (100, 200, 300, and 400 kg ha^{-1} year^{-1}) are applied as per Eq. (1.17)? If not, what is the trend of proportion of N_2O emission to fertilizers applied at different levels?

Answer (No, N_2O emission will increase polynomials, and proportion of N_2O emission to N level will increase linearly with increased N level)

7. Calculate CO_2 emission in rice crop by using Eq. (1.18), if SOC is 0.3%, bulk density is 1.55 g cm^{-3}, soil depth is 30 cm, rice duration of 110 days, average temperature is 34°C, and technology-specific index is 1. Also rank the soil factors influencing emission of CO_2 in increasing order by varying the soil data (depth, organic carbon, and bulk density) and keeping the crop duration and average temperature same.

Answer (486.8 kg CO_2 ha^{-1}, depth = OC > bulk density)

8. Calculate CO_2 emission from different farm operations (with number of operation in bracket) of mouldboard plowing (1) + cultivation (2) + sowing (1) + fertilizer application (2) + biocide use (100 g) + irrigation (400 mm) + manure use (1000 kg) + harvesting (1) by using the coefficient given in Table 1.3 and the following equation:

$$CO_2 em\, op \left(kg\ C\ ha^{-1}\right) = Moulboard\ ploughing \times 15.2 + Cultivation \times 4$$
$$+ Sowing \times 3.2 + Fertilizer\ application \times 7.6$$
$$+ Biocide\ use \times 1.4 + Irrigation\ of\ 400\,mm \times 5$$
$$+ Manure\ use \times 3.2 + Harvesting \times 10$$

Answer (5928 kg CO_2 ha^{-1})

9. Calculate indigenous and indigenous and actual methane emission using Eqs. (1.19), (1.20) from the rice crop with the data of SOC = 0.3%, bulk density = 1.55 g cm^{-3}, soil depth = 30 cm, rice duration = 110 days, and average temperature = 34°C.

Answer (72.3 kg CO_2 ha^{-1}; 90.4 kg CO_2 ha^{-1})

10. Calculate N_2O emission using Eq. (1.21).

Answer (130.5 kg C ha^{-1})

11. Calculate global warming potential using Eq. (1.22) and answers of questions 7–10.

Answer (68093 kg CO_2 ha^{-1})

12. Arrange the following fertilizers from higher to lower order of N_2O emissions:
Synthetic fertilizers, liquid organic fertilizers, unfertilized, organic-synthetic mixtures, and solid organic fertilizers

Answer (highest from liquid organic fertilizers followed by organic-synthetic mixtures and synthetic fertilizers and lowest in solid organic fertilizers and unfertilized).

13. Arrange the following moisture regimes in decreasing order for methane emission from various rice ecosystems like deep water table, irrigated continuous flooding, irrigated intermittent flooding, and rain-fed:

Answer (irrigated continuous flooding > rain-fed > irrigated intermittent flooding > deep water)

FILL IN THE BLANKS

i. In general, the materials with a C:N ratio >25:1 stimulate_____ of N, while those with a C:N ratio <25:1 stimulate _____of N.

Answer (immobilization, mineralization)

ii. The N_2O fluxes are low and slower from urea than ammonium nitrate because from urea_____.

Answer (NH_4^+ has to be hydrolyzed before being available for nitrification and denitrification processes)

iii. Nitrification is normally low below _____% water-filled pore space and release of both N_2O and N_2 due to promoted denitrification occurs above _____% water-filled pore space.

Answer (40 and 60–70)

iv. Emission of N_2O increased with increase in soil temperature from ____°C to _____°C.

Answer (5 and 40)

v. The N_2O is released by nitrification process, when base of fertilizer is _____ and soil is well aerated and yet moist, and by denitrification, when base of fertilizer is_____ and soil aeration is restricted.

Answer *(NH_3 and NO_3)*

vi. With the addition of organic materials, $N_2O:N_2$ ratio is _____, but the amount of N_2O produced is _____ from denitrification.

Answer *(decreased and increased)*

vii. In soil, CO_2 is produced through autotrophic or heterotrophic _____ of carbon organic compounds and _____of inorganic carbonate and _____of soil organic carbon.

Answer *(oxidation, chemical weathering, and mineralization)*

viii. In agriculture, CO_2 is also released through _____ and _____ combustion of fuel through the use of N, P, and K fertilizer; pesticides; and different farm operations.

Answer *(direct and indirect)*

ix. Tillage _____ the emission of CO_2 more in finer- than coarser-textured soils.

Answer (*enhances*)

x. CO_2 emission is _____ with deforestation in both the cases whether the system is restored for cultivation or left as such.

Answer (*increased*)

xi. In fine-textured soil, net mineralization of SOC and decomposition gets restricted due to _____.

Answer (*greater physical protection against microbial biomass attack*)

xii. Ammonium sulfate reduces CH_4 because of_____, and potassium nitrate increases CH_4 emission by _____.

Answer (*the competition of sulfate-reducing bacteria with methanogens and increased redox potential by adding the NO_3^-*)

xiii. Under no-tillage system, CH_4 oxidation is _____with stubble retention and is _____with no stubble retention.

Answer (*lower and enhanced*)

REFERENCES

Archibeque, S. L., Burns, J. C., & Huntington, G. B. (2001). Urea flux in beef steers: Effects of forage species and nitrogen fertilization. *Journal of Animal Science, 79*, 1937–1943.

Archibeque, S. L., Burns, J. C., & Huntington, G. B. (2002). Nitrogen metabolism of beef steers fed endophyte-free tall fescue hay: Effects of ruminally protected methionine supplementation. *Journal of Animal Science, 80*, 1344–1351.

Baldioli, M., Engel, T., Klocking, B., Priesack, E., Schaaf, T., Sperr, C., et al. (1994). *Expert-N.ein Baukastern Zur Simulation der Stickstoffdynamik in boden und Pflanze*. Prototype, Benutzerrrhandbuch, Lehreinheit fur Ackerbau und Informatik im Pflanzenbau (pp. 1–106). Freising: TU Munchen.

Battle, M., Bender, M., Sowers, T., Tans, P. P., Butler, J. H., Elkins, J. W., et al. (1996). Atmospheric gas concentrations over the past century measured in air form firm at the South Pole. *Nature, 383*, 231–235.

Bhatia, A., Jain, N., & Pathak, H. (2013). Methane and nitrous oxide emissions from Indian rice paddies, agricultural soils and crop residue burning. *Greenhouse Gases: Science and Technology, 3*, 1–16.

Bhatia, A., Pathak, H., & Aggarwal, P. K. (2004). Inventory of methane and nitrous oxide emissions from agricultural soils of India and their global warming potential. *Current Science, 87*, 317–324.

Biswas, W., Barton, L., & Carter, D. (2008). Global warming potential of wheat production in Western Australia: A life cycle assessment. *Water and Environment Journal, 22*, 206–216.

Blackmer, A. M., Bremner, J. M., & Schmidt, E. L. (1980). Production of nitrous oxide by ammonia-oxidizing chemoautotrophic microorganisms in soil. *Applied and Environmental Microbiology, 40*, 1060–1066.

Blackmer, T. M., Schepers, J. S., Varvel, G. E., & Walter-Shea, E. A. (1996). Nitrogen deficiency detection using reflected short-wave radiation from irrigated corn canopies. *Agronomy Journal, 88*, 1–5.

Bodelier, P. L. E., Roslev, P., Henckel, T., & Frenzel, P. (2000). Stimulation by ammonium-based fertilizers of methane oxidation in soil around rice roots. *Nature, 403*, 421–424.

Bouwman, A. F. (1996). Direct emissions of nitrous oxide from agricultural soils. *Nutrient Cycling in Agroecosystems, 46,* 53–70.

Butler, J. H., & Stephan, A. M. (2016). *The NOAA annual greenhouse gas index (AGGI).* Earth System Research Laboratory, R/GMD, 325 Broadway, Boulder, CO 80305–3328 http://esrl.noaa.gov/gmd//ccgg/index.html.

Campbell, C. A. (1978). Soil organic carbon, nitrogen and fertility. In M. Schnitzer & S. U. Khan (Eds.), *Soil organic matter* (pp. 173–271). Amsterdam, Netherlands: Elsevier Science.

Cartar, M. R. (1996). Analysis of soil organic matter in agroecosystems. In M. R. Carter & B. A. Stewart (Eds.), *Structure and organic matter storage in agricultural soils* (pp. 3–11). Boca Ration, FL: CRC Press, Inc.

Chadwick, D. R., Pain, B. F., & Brookman, S. K. E. (2000). Nitrous oxide and methane emissions following application of animal manures to grassland. *Journal of Environmental Quality, 29,* 277–287.

Chapman, S. J., & Thurlow, M. (1998). Peat respiration at low temperatures. *Soil Biology and Biochemistry, 30,* 1013–1021.

Christensen, B.T. (1996). Carbon in primary and secondary organo-mineral complexes. In M.R. Carter, & B.A. Stewart (Eds.), Structure and organic matter storage in agricultural soils. (pp. 97–165). Boca Raton, FL: CRC Press Inc.

Cole, C. V., Duxbury, J., Freney, J., Heinemeyer, O., Minami, K., Mosier, A., et al. (1997). Global estimates of potential mitigation of greenhouse gas emissions by agriculture. *Nutrient Cycling in Agroecosystems, 49,* 221–228.

Dalal, R. C., & Mayer, R. J. (1986). Long-term trends in fertility of soils under continuous cultivation and cereal cropping in southern Queensland. 2. Total organic C and its rate of loss from the soil profile. *Australian Journal of Soil Research, 24,* 281–292.

Dalal, R. C., Wang, W., Robertson, G. P., & Parton, W. J. (2003). Nitrous oxide emission from Australian lands and mitigation options: A review. *Australian Journal of Soil Research, 41,* 165–195.

Davidson, E. A., & David, K. (2014). Inventories and scenario of nitrous oxide emissions. *Environmental Research Letters, 9,* 1–12.

De la Chesnaye, F. C., Delhotal, C., DeAngelo, B., Ottinger Schaefer, D., & Godwin, D. (2006). *Past, present, and future of non-CO_2 gas mitigation analysis in human-induced climate change: An interdisciplinary assessment.* Cambridge: Cambridge University Press.

Dean, J.D., Huyakorn, P.S., Donigian, A.S., Voos, K.A., Schanz, R.W., & Carsel, R.F. (1989). Risk of unsaturated/saturated transport and transformation of chemical concentrations (RUSTIC). Volume II: User's guide. United States Environmental Protection Agency, Environmental Research Laboratory, EPA/600/3–89/048b, pp. 355.

Denman, K. L., Brasseur, G., Chidthaisong, A., Ciais, P., Cox, P. M., Dickinson, R. E., et al. (2007). Couplings between changes in the climate system and biogeochemistry. In: S. Solomon, D. Qin, M. Manning, Z. Chen, M. Marquis, & K. B. Averyt et al. *Climate Change 2007: The Physical Science Basis Contribution of Working Group I to the Fourth Assessment Report of the Intergovernmental Panel on Climate Change,* Cambridge, New York: Cambridge University Press.

Denmead, O. T., Freney, J. R., & Simpson, J. R. (1979). Nitrous oxide emission during denitrification in a flooded field. *Soil Science Society of America Journal, 43,* 716–718.

Eduardo, A., Luis, L., Alberto, S. -C., Josette, G., & Antonio, V. (2013). The potential of organic fertilizers and water management to reduce N_2O emissions in Mediterranean climate cropping system: A review. *Agriculture Ecosystems and Environment, 164,* 32–52.

Engel, T., & Priesack, E. (1993). Expert-N, a building block system of nitrogen models as a resource for advice, research, water management and policy. In H. J. P. Eijsackers & T. Hamers (Eds.), *Integrated soil and sediment research: A basis for proper protection* (pp. 503–507). Dordrecht: Kluwer.

Eswaran, H., van den Berg, E., & Reich, P. (1993). Organic carbon in soils of the world. *Soil Science Society of America Journal, 57*, 192–194.

Falloon, P., Smith, P., Coleman, K., & Marshall, S. (1998). Estimating the size of the inert organic matter pool for use in the Rothamsted carbon model. *Soil Biology and Biochemistry, 30*, 1207–1211.

FAO. (2003). *Trade reforms and food security: Conceptualizing the linkages*. Rome: FAO.

Firestone, M. K., & Davidson, E. A. (1989). Microbial basis of NO and N_2O production and consumption in soils. In M. Andreae & O. D. S. Schimel (Eds.), *Exchange of trace gases between terrestrial ecosystems and atmosphere* (pp. 7–21). New York: John Wiley and Sons.

Firestone, M. K., Firestone, R. B., & Tiedje, J. M. (1980). Nitrous oxide from soil denitrification: Factors controlling its biological production. *Science, 208*, 749–751.

Galloway, J. N., Schlesinger, W. H., Levy, H., Michaels, A., & Schnoor, J. L. (1995). Nitrogen fixation: Anthropogenic enhancement-environmental response. *Global Biogeochemical Science, 9*, 235–252.

Garnier, J., Billen, G., Vilain, G., Martinez, A., Silvestre, M., Mounier, E., et al. (2009). Nitrous oxide (N_2O) in the Seine River and basin: Observations and budgets. *Agriculture Ecosystems and Environment, 133*, 223–233.

GOI. (2015). *Statistics related to climate change—India 2015*. New Delhi: Ministry of Statistics and Programme Implementation, Central Statistics Office Social Statistics Division.

Goldberg, J., Trewick, S. A., & Paterson, A. M. (2008). Evolution of New Zealand's terrestrial fauna: A review of molecular evidence. *Philosophical Transactions of the Royal Society of London B: Biological Sciences, 63*, 3319–3334.

Granli, T., & Bockman, O. C. (1994). Nitrous oxide from agriculture. *Norwegian Journal Agricultural Science Supplement, 12*, 7–128.

Hendriks, D. M. D., Dolman, A. J., van der Molen, M. K., & van Huissteden, J. (2008). A compact and stable eddy covariance set-up for methane measurements using off-axis integrated cavity output spectroscopy. *Atmospheric Chemistry and Physics, 8*, 431–443.

Hoben, J. P., Gehl, R. J., Millar, N., Grace, P. R., & Robertson, G. P. (2011). Nonlinear nitrous oxide (N_2O) response to nitrogen fertilizer in on-farm corn crops of the US Midwest. *Global Change Biology, 17*, 1140–1152.

International energy agency (IEA). (2006). http://www.iea.org/statistics.

IPCC (2001). In J. T. Houhton, Y. Ding, D. J. Griggs, M. Noguer, P. J. van der Linden, & X. Dai, et al. (Eds.), *Technical summary in climate change, 2001: The scientific basis. Contribution of Working Group I of Intergovernmental Panel on Climate Change*. Cambridge: Cambridge University Press.

IPCC (2006). *Guidelines for national greenhouse gas inventories. Agriculture, Forestry and Other Land Use: (Vol. 4)*. Japan: Intergovernmental Panel on Climate Change, IGES.

Jain, A.K. (2000). The web interface of integrated science model (ISAM). http://frodo.atmos.uiuc.edu/isam.

Jarvis, P., Rey, A., Petsikos, C., Wingate, L., Rayment, M., Pereira, J., et al. (2007). Drying and wetting of Mediterranean soils stimulates decomposition and carbon dioxide emission: The Birch effect. *Tree Physiology, 27*, 929–940.

Jenkinson, D. S., & Rayner, J. H. (1977). The turnover of soil organic matter in some of the Rothamsted classical experiments. *Soil Science, 123*, 298–305.

Kasterine, A., & Vanzetti, D. (2010). *The effectiveness, efficiency and equity of market-based instruments to mitigate GHG emission from the agri-food sector*. UNCTAD Trade and Environment Review 2009/2010, Geneva. www.unctad.org/Templates/WebFlyer. asp?intItemID=5304&lang=1.

Kebreab, E., Clark, K., Wagner-Riddle, C., & France, J. (2006). Methane and nitrous oxide emissions from Canadian animal agriculture: A review. *Canadian Journal of Animal Science, 86*, 135–158.

Kim, D. G., Hernandez-Ramirez, G., & Giltrap, D. (2013). Linear and nonlinear dependency of direct nitrous oxide emissions on fertilizer nitrogen input: A meta-analysis. *Agriculture, Ecosystems and Environment, 168*, 53–65.

Knoblauch, C., Zimmermann, U., Blumenberg, M., Michaelis, W., & Pfeiffer, E. M. (2008). Methane turnover and temperature response of methane-oxidizing bacteria in permafrost-affected soils of northeast Siberia. *Soil Biology and Biochemistry, 40*, 3004–3013.

Lal, R. (1995). Global soil erosion by water and carbon dynamics. In R. Lal, J.K. Kimble, E, Levine, & B.A. Stewart (Eds.), Soils and global change (pp. 131–141). Boca Raton, FL: CRC/Lewis Publishers.

Lal, R. (2001). Soil degradation by erosion. *Land Degradation and Development, 12*, 519–539.

Lal, R. (2004). Carbon emission from farm operations. *Environmental International, 30*, 981–990.

Lambert, M. G., Devantler, B. P., Nesip, B., & Penny, E. (1985). Losses of nitrogen, phosphorus and sediment in runoff from hill country under different fertilizer and grazing management regimes. *New Zealand Journal of Agricultural Research, 28*, 371–379.

Laverman, A. M., Garnier, J. A., Mounier, E. M., & Roose-Amsaleg, C. L. (2010). Nitrous oxide production kinetics during nitrate reduction in river sediments. *Water Research, 44*, 1753–1764.

Li, C., Frokling, S., & Frokling, T. A. (1992). A model of nitrous oxide evolution from soil driven by rain events: 1. Model structure and sensitivity. *Journal of Geophysics Research, 97*, 9759–9776.

Li, C., Mosier, A., Wassmann, R., Cai, Z., Zheng, X., Huang, Y., et al. (2004). Modeling greenhouse gas emissions from rice-based production systems: Sensitivity and upscaling. *Global Biogeochemical Cycles, 18*, 1–19.

Meyer, C. P., Galbally, I. E., Wang, Y., Weeks, I. A., Jamie, I., & Griffith, D. W. T. (2001). *Two automatic chamber techniques for measuring soil-atmosphere exchange of trace gases and results of their use in the oasis field experiments.* CSIRO Atmosphere Research Technical Paper No 51 (pp. 1–33).

Millar, N., Robertson, G. P., Diamant, A., Gehl, R. J., Grace, P. R., & Hoben, J. P. (2013). Quantifying N_2O emissions reduction in US agricultural crops through N fertilizer rate reduction. www.v-c-s.org/methodologies/quantifying-n2o-emissions-reductions-agricultural-crops-through-nitrogen-fertilizer.

Mohanty, S. R., Nayak, D. R., Babu, Y. J., & Adhya, T. K. (2004). Butachlor inhibits production and oxidation of methane in tropical rice soils under flooded condition. *Microbiology Research, 159*, 193–201.

Mosier, A. R. (1998). Chamber and Isotope techniques. In M. O. Andreae & D. S. Schimel (Eds.), *Exchange of trace gases between terrestrial ecosystems and the atmosphere* (pp. 175–187). New York: John Wiley and Sons.

NOAA. (2008). *National ocean and atmospheric administration. In B. M. Tignor & H. L. Miller (Eds.), Global climate change impacts in United States.* Cambridge, New York: Cambridge University Press. www.noaa.govt.

Parton, W. J., Hartman, M. D., Ojima, D. S., & Schimel, D. (1998). DAYCENT: Its land surface submodel: Description and testing. *Global and Planetary Change Journal, 19*, 35–48.

Parton, W. J., Mosier, A. R., Ojima, D. S., Valentine, D. W., Schimel, D. S., Weier, K., et al. (1996). Generalized model for N_2 and N_2O production from nitrification and denitrification. *Global Biogeochemical Cycles, 10*, 401–412.

Parton, W. J., Schimel, D. S., Cole, C. V., & Ojima, D. S. (1994). A generalized model for soil organic matter dynamics: Sensitivity to litter chemistry, texture and management. In R. B. Bryanti & R. W. Arnold (Eds.), *Quantitative modelling of soil forming processes* (pp. 147–167). Soil Science Society of America Journal, Special Publication 39, Madison, WI.

Parton, W. J., Schimel, D. S., Cole, C. V., & Ojirna, D. S. (1987). Analysis of factors controlling soil organic matter levels in Great Plains grasslands. *Soil Science Society of America Journal, 51,* 1173–1179.

Pathak, H. (1999). Emissions of nitrous oxide from soil. *Current Science, 77,* 359–369.

Pathak, H., Agarwal, T., & Jain, N. (2012). Greenhouse gas emission from rice and wheat systems: A life-cycle assessment. In H. Pathak & P. K. Aggarwal (Eds.), *Low carbon technologies for agriculture: A study on rice and wheat systems in the Indo-Gangetic Plains.* New Delhi: Indian Agricultural Research Institute.

Pathak, H., Upadhyay, R. C., Muralidhar, M., Bhattacharyya, P., & Venkateswarlu, B. (2013). Measurement of greenhouse gas emission from crop, livestock and aquaculture. In *NICRA Manual Series 2.* New Delhi: Indian Agricultural Research Institute.

Pathak, H., & Wassmann, R. (2007). Introducing greenhouse gas mitigation as a development objective in rice-based agriculture: I. Generation of technical coefficients. *Agricultural Systems, 94,* 807–825.

Potter, C. S., Riley, R. H., & Klooster, S. A. (1997). Simulation modeling of nitrogen traces gas emissions along an age gradient of tropical forest soils. *Ecological Modelling, 97,* 179–196.

Pretty, J. N., Ball, A. S., Xiaoyun, L., & Ravindranathan, N. H. (2002). The role of sustainable agriculture and renewable-resource management in reducing greenhouse-gas emissions and increasing sinks in China and India. *Philosophical Transactions of the Royal Society A, 360,* 1741–1761.

Rath, A. K., Swaina, B., Ramakrishnan, B., Panda, D., Adhya, T. K., Rao, V. R., et al. (1999). Influence of fertilizer management and water regime on methane emission from rice fields. *Agriculture, Ecosystems and Environment, 76,* 99–107.

Riley, W.J., & Matson, P.A. (1989). The NLOSS model. 1998 Fall AGU meeting. Poster A41B-25. American Geophysical Union, 79.

Ruan, L. (2014). Impacts of biofuel crops on greenhouse gas emissions from agricultural ecosystems. (PhD dissertation), Michigan State University, East Lansing, MI.

Saggar, S., Bolan, N. S., Bhandral, R., Hedley, C. B., & Luo, J. (2004). A review of emissions of methane, ammonia, and nitrous oxide from animal excreta deposition and farm effluent application in grazed pastures. *New Zealand Journal of Agricultural Research, 47,* 513–544.

Saggar, S., Luo, J., Giltrap, D. L., & Maddenna, M. (2009). Nitrous oxide emission from temperate grasslands, processes, measurements, modelling and mitigation. In I. S. Adam & E. P. Barnhart (Eds.), *Nitrous oxide emissions research progress* (pp. 1–66). New York: Nova Science Publishers Inc.

Schlesinger, W. H. (1999). Carbon sequestration in soils. *Science, 284,* 2095.

Shcherbak, I., Millar, N., & Robertson, G. P. (2014). Global metaanalysis of the nonlinear response of soil nitrous oxide (N_2O) emissions to fertilizer nitrogen. *Proceedings of the National Academy of Sciences, 111,* 9199–9204.

Shindell, D. T., Walter, B. P., & Faluvegi, G. (2004). Impacts of climate change on methane emissions from wetlands. *Geophysics Research Letter, 31,* L21202.

Shine, K. P., Derwent, R. G., Wuebbles, D. J., & Morcrette, J. J. (1990). Radiative forcing of climate. In J. T. Houghton, G. J. Jenkins, & J. J. Ephraums (Eds.), *Climate Change: The IPCC Scientific Assessment* (pp. 41–68). Cambridge: Cambridge University Press.

Singh, S., Singh, S., Pannu, C. J. S., & Singh, J. (1999). Energy input and yield relations for wheat in different agro-climatic zones of the Punjab. *Applied Energy, 63,* 287–298.

Smith, P., Martino, D., Cai, Z., Gwary, D., Janzen, H. H., Kumar, P., et al. (2007). Agriculture. In B. Metz, O. R. Davidson, P. R. Bosch, R. Dave, & L. A. Meyer (Eds.), *Agriculture. Chapter 8 of Climate Change: Mitigation.* Cambridge, New York: Cambridge University Press. Contribution of working group III to the intergovernmental Panel on Climate Change. Fourth assessment report.

Suwanwaree, P., & Robertson, G. P. (2005). Methane oxidation in forest, successional, and no-till agricultural ecosystems: Effects of nitrogen and soil disturbance. *Soil Science Society of America Journal, 69,* 1722–1729.

Tejada, M., & Gonzalez, J. L. (2006). The relationships between erodibility and erosion in a soil treated with two organic amendments. *Soil and Tillage Research, 91,* 186–198.

Thomson, A. J., Giannopoulos, G., Pretty, J., Baggs, E. M., & Richardson, D. J. (2012). Biological sources and sinks of nitrous oxide and strategies to mitigate emissions. *Philosophical Transactions of the Royal Society Biological Sciences, London, 367,* 1157–1168.

Tirado, R., Gopikrishna, S. R., Krishnan, R., & Smith, P. (2010). Greenhouse gas emissions and mitigation potential from fertilizer manufacture and application in India. *International Journal of Agricultural Sustainability, 8,* 176–185.

US-EPA (2006). *Global anthropogenic non-CO_2 greenhouse gas emissions: 1990–2020.* http://www.epa.gov/nonco2/econ-inv/international.html.

Ussiri, D. A. N., Lal, R., & Jarecki, M. K. (2009). Nitrous oxide and methane emissions from long-term tillage under a continuous corn cropping system in Ohio. *Soil and Tillage Research, 104,* 247–253.

Van Groenigen, K. J., Osenberg, C. W., & Hungate, B. A. (2011). Increased soil emissions of potent greenhouse gases under increased atmospheric CO_2. *Nature, 475,* 214–216.

Vinther, F. P., & Hansen, S. (2004). *SimDen—a simple empirical model for quantification of N_2O emission and denitrification.* DIAS Report 104 (in Danish with English summary) (pp. 1–47).

Ward, B. B. (2011). Nitrification in the ocean. In B. B. Ward, D. J. Arp, & M. G. Klotz (Eds.), *Nitrification* (pp. 325–345). Washington, DC: ASM Press.

Wassmann, R., Lantin, R. S., Neue, H. U., Buendia, L. V., Corton, T. M., & Lu, Y. (2000). Characterization of methane emissions from rice fields in Asia. III. Mitigation options and future research needs. *Nutrient Cycling in Agroecosystems, 58,* 23–36.

Watson, R. T., Rodhe, H., Oeschger, H., & Siegenthaler, U. (1990). Greenhouse gases and aerosols. In J. T. Houghton, G. J. Jenkins, & J. J. Ephraums (Eds.), *Climate change: The IPCC scientific assessment* (pp. 1–40). Cambridge: Cambridge University Press.

Watson, R. T., Zinyowera, M. C., & Moss, R. H. (1996). Technical summary: Impacts, adaptations and mitigation options. In R. T. Watson, M. C. Zinyowera, R. H. Moss, & D. J. Dokken (Eds.), *Impacts, adaptations and mitigation of climate change: Scientific-technical analyses* (pp. 19–53). Cambridge: Cambridge University Press.

FURTHER READING

Bern climate change model. www.ipcc-data.org/ancillary/tar-bern.txi.

Eswaran, H., van den Berg, E., Reich, P., & Kimble, J. (1995). Global soil carbon resources. In R. Lal, J. Kimble, E. Levine, & B. A. Stewart (Eds.), *Soils and global change* (pp. 27–43). Boca Raton, FL: Lewis Publishers.

CHAPTER TWO

Climate Change Projections

Contents

2.1 INTRODUCTION

Atmospheric conditions experienced by the living beings around them are often expressed as weather and climate vaguely. To understand climate change and its projections, it is obligatory that one should be able to differentiate between weather and climate, climate variability, and climate change. By definition, *weather* is a state of the atmosphere at a given time. The state of atmosphere is usually determined in terms of its temperature, atmospheric composition (e.g., water vapors or carbon dioxide content), wind speed and direction, precipitation (P_{cp}), pressure, and density in addition to the intensity of solar and terrestrial emitted radiation. The weather is a primarily determinant of agricultural production. *Climate* is the average weather over time and is depicted in terms of mean (μ) and variance (σ^2) of the key weather parameters for a given period, usually 30 years. Variation in climate parameters for a specific duration having the same μ value is called *climate variability*. The change in climate over time when μ and other statistical parameters are also changed is called *climate change*. The climate change occurs due to the interaction between the constituents of climate system such as atmosphere and underlying surface-ocean, land, and ice on the earth surface. The climate change is assessed from the past observed data and is projected with the help of climate models.

The climate models often used are global climate models, also known as general circulation models (GCMs), which simulate past weather and generate future modeled data on temperature, P_{cp}, and wind depending partially on the atmospheric concentration of greenhouse gases (GHGs), derived from future scenarios and on the model simulation. Each run of the model is different as weather is partly a stochastic phenomenon. The GCMs are with coarse spatial resolution (typically around 100–300 km) and have a number of uncertainties caused by initial and boundary conditions, observations, model parameters, and structure. Due to these uncertainties, some processes in the climate system remain ambiguous. Such uncertainties can be minimized by using ensemble simulations wherein different models are averaged together. The GCMs have large variability both at spatial and temporal scale and are not apposite to evaluate climate change impacts on agricultural systems due to local topographic features. For assessing possible impacts on agriculture at regional level and incorporation of the local features, the projections from GCMs have to undergo downscaling in the form of either statistical downscaling or dynamic downscaling. However, the downscaled data may have inconsistency in magnitude and time trends, which needs application of bias correction or change methods to confine the gap between observed and modeled data.

The purpose of writing this chapter is to illustrate GCMs, their uncertainties, and ways to reduce those for better understanding climate change projections. This chapter provides a detailed description of different emission scenarios from Special Report on Emission Scenario (SRES) to Representative Concentration Pathways (RCPs). It also covers the downscaling of GCMs to regional climate model (RCM) data (by statistical and dynamical) and bias correction methods to minimize bias in the projected climate data to have more appropriate estimates of the climate change. Trends and magnitudes of climate change in the past and their projections at global, regional, and local scale are also conversed.

2.2 EMISSION SCENARIOS

Emission scenarios describe the concentrations of GHGs, aerosols, and other pollutants into the atmosphere from various sources, that is, natural and man-made to which climate is sensitive, in addition to description on land cover and land use. By definition, scenario is a logical and reasonable depiction of a likely future state of the climate at global level. Scenario is also addressed by other synonyms such as characterization, story line, and construction. Alike, demographic, politicosocietal, and technological story lines illustrate a *scenario family*. A future climate scenario is generated by amending the baseline data with difference or proportion between mean outputs from GCM control simulations (usually 10–30 years) and projected GCM mean outputs of the future. Difference is usually applied for temperature and ratio for P_{cp}. Over time, a variety of approaches, that is, first assessment report, 1990 (SA90); second assessment report, 1992 (IS92); third and fourth assessment reports, 2000 (SRES); and fifth assessment report, 2009 (RCP), to emission scenarios have been used in climate research (Bjornaes, 2013). Presently, both SRES and RCP scenarios are being used by the research workers for projecting future climate and prioritizing adaptation strategies.

2.2.1 Special Report on Emission Scenarios

There are 40 different scenarios developed by six modeling teams, each one with diverse hypothesis for future GHG levels, land use, and additional responsible factors. IPCC SRES team gave four narrative story lines, namely, A1, A2, B1, and B2, that depicted the association between the driving forces of GHGs and aerosol emissions and their progression during the 21st century globally (IPCC, 2000). In general, the four story lines describe two sets of opposing tendencies: one set varying between strong economic (A1) and strong environmental (B1) values and the other set between increasing globalization (A2) and increasing regionalization (B2). Every story line represents different demographic, social, economic, technological, and environmental developments that diverge in increasingly irreversible ways.

2.2.2 Representative Concentration Pathways

More recent of SRES scenarios are RCP (Weyant et al., 2009). These are reliable sets of projections of the components of radiative forcing (defined as *the alteration in the balance between incoming and outgoing radiation to the atmosphere and are mainly due to changes in atmospheric composition*), which are intended to be provided as input for climate modeling. Radiative forcing, measured in watts per square meter (Wm^{-2}), is the extra heat that the lower atmosphere will retain because of the presence of additional GHGs and aerosols. The RCPs are time- and space-dependent trajectories of concentrations of GHGs and pollutants resulting from human activities including changes in land use. The RCPs cover a wider range of possibilities (Table 2.1) than the

Table 2.1 Overview of Representative Concentration Pathways (RCPs)

RCP 8.5	Rising radiative forcing pathway leading to $8.5\,Wm^{-2}$ in 2100	• Three times today's CO_2 emissions by 2100 • Rapid increase in methane emissions • Increased use of croplands and grassland, which is driven by an increase in population • A world population of 12 billion by 2100 • Lower rate of technology development • Heavy reliance on fossil fuels • High energy intensity • No implementation of climate policies
RCP 6.0	Stabilization without overshoot pathway to $6\,Wm^{-2}$ at stabilization after 2100	• Heavy reliance on fossil fuels • Intermediate energy intensity • Increasing use of croplands and declining use of grasslands • Stable methane emissions • CO_2 emissions peak in 2060 at 75% above today's levels and then decline to 25% above today's levels
RCP 4.5	Stabilization without overshoot pathway to $4.5\,Wm^{-2}$ at stabilization after 2100	• Lower energy intensity • Strong reforestation programs • Decreasing use of croplands and grasslands due to yield increase and dietary changes • Stringent climate policies • Stable methane emissions • CO_2 emissions increase only slightly before decline commences around 2040

Table 2.1 Overview of Representative Concentration Pathways (RCPs)—cont'd

RCP 2.6	Peak in radiative forcing at $\sim 3\,\mathrm{Wm}^{-2}$ before 2100 and decline	• Declining use of oil • Low energy intensity • A world population of 9 billion by year 2100 • The use of croplands increases due to bio-energy production • More intensive animal husbandry • Methane emissions reduced by 40% • CO_2 emissions stay at today's level until 2020, then decline, and become negative in 2100 • CO_2 concentrations peak around 2050, followed by a modest decline to around 400 ppm by 2100

Modified from Bjoraes, C. (2013). A guide to representative concentration pathways (pp. 1–5). Norway: CICERO, Center for International Climate and Environmental Research.

SRES emission scenarios. There are four RCPs ranging from 2.6 to $8.5\,\mathrm{Wm}^{-2}$ and are described as follows:

RCP 8.5: This corresponds to a high-emission scenario. It represents a future with no specific plan to lessen emissions. It was synthesized by the International Institute for Applied System Analysis in Austria and is categorized by rising GHG emissions that direct to high GHG concentrations over time. Comparable SRES scenario, A1 F1.

RCP 6.0: This corresponds to intermediate level of emissions and was developed by the National Institute for Environmental Studies in Japan. In this, the radiative forcing will be stabilized shortly after year 2100, due to various technological and strategic interventions for limiting GHG emissions. Comparable SRES scenario, B2.

RCP 4.5: This RCP was developed by the Pacific Northwest National Laboratory in the United States, and it also corresponds to intermediate level of emissions. Here, radiative forcing is stabilized shortly after year 2100, consistent with a future with relatively ambitious emission reductions. Comparable SRES scenario, B1.

RCP 2.6: This corresponds to a low-emission scenario. This RCP was developed by PBL Netherlands Environmental Assessment Agency. Here, radiative forcing reaches $3.1\,\mathrm{Wm}^{-2}$ before it returns to $2.6\,\mathrm{Wm}^{-2}$ by 2100. To attain these levels, determined GHG emission cutbacks would be needed. Comparable SRES scenario, none.

Nowadays climate modelers are using both the time series of emissions and concentrations associated with the four RCPs, along with the SRES for performing climate model studies.

2.3 GLOBAL CLIMATE MODELS

A GCM aims to describe climate behavior by incorporating equations, which are either derived directly from physical laws (e.g., Newton's law) or constructed by more empirical means. Climate model differs from the weather model. Climate model is to be run with grid points significantly further apart than weather model. Climate models have undergone continuous development for the last three decades; now include interactions between the atmosphere, oceans, and land surface; and vary from simple energy-balance models (EBMs) to complicated earth system models (ESMs) that necessitate high-performance computing facilities.

2.3.1 Simple Models

Energy balance and box models: The EBMs include only a few parameters, that is, values of natural (e.g., solar) and anthropogenic forcing and some constants, which lead to an output of temperature and associated heat fluxes. These models have poor spatial resolution and are mostly one or zero dimensional. This simplicity, however, makes them quick to run. To include more dimensions, the EBMs can be expanded, and those are called box-diffusion models, which deal with interactions of different systems (such as ocean, land, and atmosphere) represented as boxes interconnected through energy transfer equations with each system (box) containing their own independent characteristics. Box models parameterize many events and still do not permit very detailed questions related to climatic mechanisms, which are considered a disadvantage.

2.3.2 Complex Models

General circulation models: The GCMs are much more comprehensive than EBMs, including more physical constituents, represented in a more complex manner. In contrast to EBMs, they don't conceptualize the world as a dot, a line, or parts of it as boxes that are interacting but incorporate the three-dimensional nature of ocean and atmosphere. These models are used to replicate the pattern of the atmosphere, oceans, land surface, and snow–covered areas over the globe. Parameters such as temperature, wind, and P_{cp} are estimated over a three-dimensional arrangement of grid cells spaced typically 250–600 km apart, 10–20 vertical layers through the height of the atmosphere, depending on the model. GCMs were started off being either an oceanic or an atmospheric circulation model, which later got combined to form the atmosphere-ocean general circulation models (AOGCMs) as shown in Fig. 2.1.

Earth system models: These are advanced versions of AOGCMs and include representation of a variety of biogeochemical processes such as those involved in the carbon cycle, the sulfur cycle, or the ozone (Flato, 2011). These models offer a wide range of tools for reproducing historic and predicting future climate in response to external forcing wherein biogeochemical responses play a significant role.

Many research institutions around the world have developed and maintained their own global climate models, and the details of few of them commonly used are listed in Table 2.2.

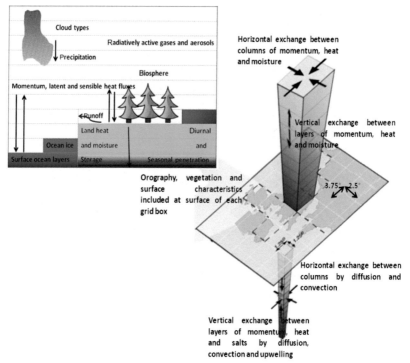

Fig. 2.1 Schematic diagrams of GCM grids. *Modified from http://www.ipcc-data.org/guidelines/pages/gcm_guide.html.*

Table 2.2 List of the Global Climate Models in IPCC AR5 (Coupled Model Intercomparison Project 5, CMIP5)

GCM	Model	Resolution	Source
BCC–CSM and BCC–CSM 1.1 (m)	It is a fully coupled global climate carbon model including interactive vegetation and global carbon cycle in which the atmospheric component, ocean component, land component, and sea ice component are fully coupled and interact with each other through fluxes of momentum, energy, water, and carbon at their interfaces. Information between the atmosphere and the ocean is exchanged once per simulated day. The exchange of atmospheric carbon with the land biosphere is calculated at each model time step (20 min). The major difference between BCC–CSM 1.1 (m) and BCC–CSM 1.1 is the model resolution in their atmospheric component. BCC–CSM 1.1 (m) has participated in the fifth phase of the Coupled Model Intercomparison Project (CMIP5) and conducted several core experiments	64×128 160×320	Beijing Climate Center, China Meteorological Administration, China
BNU-ESM	An earth system model is based on several widely evaluated climate model components and is used to study mechanisms of ocean-atmosphere interactions, natural climate variability, and carbon–climate feedbacks at interannual to interdecadal timescales	64×128	Beijing Normal University, China
CanESM2	It is the fourth-generation atmosphere–ocean general circulation model. There are 40 vertical levels with spacing ranging from 10 m near the surface (there are 16 levels in the upper 200 m) to nearly 400 m in the deep ocean. Horizontal coordinates are spherical with grid spacing ~1.41 degrees in longitude and 0.94 degrees in latitude	64×128	National Center for Atmospheric Research (NCAR), United States
CCSM4	The Community Climate System Model (CCSM) is a coupled climate model for simulating the earth's climate system, composed of four separate models simultaneously simulating the earth's atmosphere, ocean, land surface, sea ice, and one central coupler component	192×288	National Center for Atmospheric Research (NCAR), United States
CNRM-CM5	CNRM-CM5 is an Earth system model designed to run climate simulations. It consists of several existing models designed independently and coupled through the OASIS software	128×256	CERFACS (Centre National de Recherches Meteorologiques, France)
CSIRO-Mk 3.6.0	Australian Commonwealth Scientific and Industrial Research Organisation	96×192	

FGOALS-g 2	Flexible Global Ocean–Atmosphere–Land System (FGOALS) model includes four individual components (an atmosphere component, an ocean component, a land component, and a sea–ice component) that are driven by a flux coupler module	108×128	State Key Laboratory of Numerical Modeling for Atmospheric Sciences and Geophysical Fluid Dynamics, Institute of Atmospheric Physics (LASG–IAP), Chinese Academy of Sciences, Beijing, China
FIO-ESM	Earth system model, which is named as the First Institute of Oceanography-Earth System Model (FIO-ESM), is composed of a coupled physical climate model and a coupled carbon cycle model. The FIO-ESM is employed to conduct Coupled Model Intercomparison Project Phase 5 (CMIP5) experiments	64×128	The First Institute of Oceanography, SOA, China
GFDL CM3	GFDL has constructed NOAA's first Earth system models (ESMs). The atmospheric component of the ESMs includes physical features such as aerosols (both natural and anthropogenic), cloud physics, and precipitation. The land component includes precipitation, evaporation, streams, lakes, rivers, runoff, and a terrestrial ecology component to simulate dynamic reservoirs of carbon and other tracers. The oceanic component includes features such as free surface to capture wave processes; water fluxes, or flow; currents; sea-ice dynamics; iceberg transport of freshwater; a state-of-the-art representation of ocean mixing; and marine biogeochemistry and ecology	90×144	Geophysical Fluid Dynamics Laboratory, United States
GISS-E2-H and GISS-E2-R	NASA Goddard Institute for Space Studies (GISS) climate model E 2 with 2 degrees by 2.5 degrees horizontal resolution and 40 vertical layers in atmosphere, with the model top at 0.1 hPa Goddard Institute for Space Studies (NASA), United States	90×144	

Continued

Table 2.2 List of the Global Climate Models in IPCC AR5 (Coupled Model Intercomparison Project 5, CMIP5)—cont'd

GCM	Model	Resolution	Source
HadGEM2-ES	HadGEM2-ES is a coupled Earth system model that was used by the Met Office Hadley Centre for the CMIP5 centennial simulations. The HadGEM2-ES climate model comprises an atmospheric GCM at N96 and L38 horizontal and vertical resolution and an ocean GCM with a 1 degree horizontal resolution (increasing to 1/3 degrees at the equator) and 40 vertical levels. Earth system components included are the terrestrial and ocean carbon cycle and tropospheric chemistry. Terrestrial vegetation and carbon are represented by the dynamic global vegetation model, TRIFFID, which simulates the coverage and carbon balance of five vegetation types (broadleaf tree, needle leaf tree, C3 grass, C4 grass, and shrub). Ocean biology and carbonate chemistry are represented by diat-HadOCC, which includes limitation of plankton growth by macro– and micronutrients and also simulates emissions of DMS to the atmosphere	145 × 192	Met Office Hadley Center, Unites Kingdom
MPI-ESM-LR and MPI-ESMR	MPI-ESM (MPG) is a comprehensive Earth system model that consists of component models for the ocean, the atmosphere, and the land surface. These components are coupled through the exchange of energy, momentum, water, and important trace gases such as carbon dioxide. The model is developed by the MPI for Meteorology (MPI-M) and based on its predecessors, the ECHAM5/MPIOM coupled model and its COSMOS versions	96 × 192	Max Planck Institute for Meteorology (MPI-M), Germany
NorESM	The Norwegian Earth System Model (NorESM) is one out of ~20 climate models that has produced output for the NorESM1-M CMIP5 and has a horizontal resolution of ~2 degrees for the atmosphere and land components and 1 degree for the ocean and ice components. NorESM is also available in a lower-resolution version (NorESM1-L) and a version that includes prognostic biogeochemical cycling (NorESM1-ME) NorESM1-M: Norwegian Earth System Model 1—medium resolution NorESM1-ME: Norwegian Earth System Model 1—medium resolution with capability to be fully emission driven	96 × 144	

Modified from Miao, C., Duan, Q., Sun, Q., Huang, Y., Kong, D., Yang, T., et al., (2014). Assessment of CMIP5 climate models and projected temperature changes over Northern Eurasia. *Environmental Research Letters*, 9, 055007.

Even though the models are alike in several ways, still, minor differences may be present due to variation in grid characteristics, spatial resolution, parameterization schemes, model subcomponents, and climate sensitivity for different forcing factors (solar, aerosols, carbon dioxide, methane, etc.). Climate sensitivity is the steady-state increase in the global annual mean surface air temperature associated with a given global–mean radiative forcing and is mostly compared using CO_2 doubling as a benchmark.

2.4 UNCERTAINTY IN CLIMATE CHANGE MODELING

In models, the climate state changes in accordance to the set of mathematical rules based on the basic laws of material science; nevertheless, there are some uncertainties. For instance, (i) many physical processes such as those related to clouds occur at smaller scales and cannot be properly modeled over the larger scale. In such cases, the known parameters should be averaged over the larger scale through parameterization; (ii) climate models typically simulate hourly to daily weather, but for climate change, the mean weather over 30 years or more is required; (iii) a number of factors, for example, volcanic eruptions, cloud formation, water vapors, ice, ocean circulation changes over the oceans, and natural cycles of GHGs may prevent precise projection of climate change; and (iv) even though past climate changes may concur with model predictions, uncertainty may still arise owing to inaccuracies in the data and potentially significant aspect for which the complete information is not available. Even though the global climate change is largely known, it is extremely difficult to envisage in detail that how climate change will influence individual locations, especially changes in P_{cp} due to alteration in wind patterns, ocean currents, plants, and soils. Further extreme climate events and their timings cannot be foreseen with confidence. On the other hand, temperature predictions have a small degree of uncertainty than those for P_{cp} due to its less spatial variability. Such uncertainties in climate modeling hamper proper realization of the processes in climate system and are reduced with ensemble approach.

2.4.1 Ensemble Approach

This is a grouped approach, which is preferred to reduce the uncertainties arising from initial conditions. In this approach, consequences of initial conditions are minimized by performing a series of analogous model run with identical historical changes and identical future increase in GHGs, and the outputs are averaged without giving emphasis on the point of initiation on the baseline. Through this approach, the effect of initial condition becomes unimportant to the long-term change, yet year-to-year and decade-to-decade differences by natural climate variability in the resulting climate are significant, which are large at regional/local scales, especially for variables like P_{cp}.

2.4.2 Multimodel Ensemble Approach

Multimodel ensemble is used to evaluate intercomparison of GCMs. A large project called Coupled Model Intercomparison Project (CMIP3 and CMIP5) is organized by the World Climate Research Program (WCRP) of the World Meteorological Organization (WMO) to compare the results and assess the performance of the major GCMs available worldwide, in support of the work of the IPCC. Both the CMIP5 and CMIP3 GCMs are being increasingly used worldwide to make projections of future climate change. CMIP5 is included in Assessment Report 5 (AR5) and contains more and advanced models than CMIP3, which was used for Assessment Report 4 (AR4). CMIP5 uses RCPs, whereas CMIP3 uses scenarios from the IPCC SRES. The widespread variation in P_{cp} in CMIP (3 and 5) provides some insight on model uncertainty. Although multimodel experiments are not intended to provide uncertainty analysis, through systematic intercomparisons, they offer an ensemble of opportunity and help to (i) characterize uncertainty; (ii) understand the source of deviation across an ensemble; (iii) formulate the best statistical structure; and (iv) choose the correct model quality metrics. Team et al. (2010) emphasized that while evaluating outcomes of ensembles from multimodel experiments, one should consider the following aspects:

1. There should be clarity on the purpose of forming and interpreting ensembles and the knowledge of the deviations between the model setup and model simulations.
2. The distinction between "best effort" simulations (i.e., the results from the default version of a model submitted to a multimodel database) and perturbed physics ensembles must be recognized.
3. In many cases, it may be appropriate to consider simulations from CMIP3 and combine CMIP3 and CMIP5 recognizing differences in specifications (e.g., differences in forcing scenarios). This is important when the number of ensemble members or simulations differs between contributing models. In a few ensemble simulations, the members may require to be averaged first before combining different models, while in other cases, only one member may be used for each model.

4. Ensemble members may possibly not correspond to the climate system behavior (trajectory) entirely independent of one another. This is probably true in case the members are different versions of the same model or initial conditions are the same. But still, different models may share components and choices of parameterizations of processes and may have been calibrated using the same data sets.

2.5 DOWNSCALING OF GLOBAL CLIMATE MODELS

At present, GCM is the most important tool for representing future climate conditions at a coarse spatial resolution. But at local/regional scale, where processes such as mountain ranges blocking air flow or dust clouds interacting with radiation are involved, the magnitude and pattern of climate change and its impact on agricultural system may not be the same as from GCMs per se. Under such situations, the data of GCM are downscaled. Downscaling is a technique of bridging the gap between the low-spatial-resolution GCMs and the high-spatial-resolution regional-, catchment-, or point-scale data. Spatial detail in GCM-generated climate change scenarios is interpolated to a finer resolution and then combining these outputs with observed climate data of the region. For this, observed mean monthly climate data or an observed monthly or daily statistic for a site or catchment may be perturbed by the GCM changes interpolated to the site in question. This is the easiest way of downscaling and provides the climate scenarios at a resolution that would otherwise be difficult or costly to obtain.

2.5.1 Statistical Downscaling

As the climate at regional scale is governed by the global-scale variations in the climate and local/regional physiographic features (i.e., topography, land–sea distribution, and land use), finer-scale (regional) meteorologic data from GCMs are generally derived by statistical downscaling assuming that the relationship between present global and local climate circulation will be same in future climates (Zorita & Von Storch, 1999). In this approach, an empirical or statistical relationship is established from observations between large-scale variables (or predictors) like atmospheric surface pressure and a local variable (or predictands) like the wind speed at a particular site. Then, the outputs (predictors) from GCM simulations are given as input to the statistical model for estimating the corresponding local/regional climate characteristics. Predictor sets are derived from sea-level pressure, geopotential height, wind fields, absolute or relative humidity, and temperature variables (IPCC-TGICA, 2007). Statistical downscaling consists of a number of methods, which are comparatively easy to apply, but requires site-specific information on observed climate data. There are three broad clusters of statistical methods, that is, linear methods, weather classifications, and stochastic weather generators. Linear regression methods are practical to a distinct predictor-predictand pair or spatial parameters of predictors-predictands. In case of multiple variables, the predictor variable should be principal component. However, the application of this method is limited to the cases where both the predictor and the predictand values follow the normal distribution. It may fail to predict daily P_{cp} values as the distribution becomes nonnormal and unsymmetrical owing to the presence of a large number of extreme events. Linear methods include delta change, multiple regression, and linear regression (Trzaska & Schnarr, 2014). The delta change technique uses the factor, which is the obtained as difference in average value of the climate parameter in the future and the historical simulations. The factor is then applied to the observed time series to transform into a new time series corresponding of the future climate. In general, empirical method, being inexpensive, is preferred over numerical in statistical downscaling.

Weather classification methods are used to forecast the local climate depending on large-scale atmospheric climate. Weather classifications include analog methods such as cluster analysis and artificial neural network. In analog method, future atmospheric condition, simulated by a GCM, is matched with its analogous historic (observed) atmospheric condition. The selected historic atmospheric condition corresponds to a value or a class of values of the local variable, which are afterward simulated under the future atmospheric condition.

These methods can be used for downscaling nonnormal distributions such as daily P_{cp}. However, it requires roughly approximately 30 years of daily data to assess weather conditions. Moreover, these methods are computationally challenging as compared with the linear ones because a large amount of daily data have to be generated and analyzed. In cluster analysis, the data are grouped in cluster/class. This method explains the complex relationships about large- to local-scale climate by analyzing natural clusters or types. Artificial neural networks (ANNs) are largely algorithms that use stepwise nonlinear functions to transform input data sets into output data sets (Benestad, 2008).

The stochastic weather generators are largely used for temporal downscaling to provide at a daily resolution local spatial data in certain impact study models, wherein GCM data are not reliable. A stochastic weather generator uses statistical parameters (derived from actual observations) to generate artificial time series of weather conditions with unlimited length for a location. Wherein data are limited, the stochastic weather generators are also used to generate daily data of weather parameters such as P_{cp}, maximum temperature (T_{max}), minimum temperature (T_{min}), and humidity from monthly or yearly averages for a region. Weather generators can also predict several variables while preserving the coherency, for example, generation of a daily sequence of insolation (*the solar radiation reaching the earth's surface per square centimeter per minute*) corresponding to daily sequence of rainy and dry days. However, the generated time series generally ignore the spatial relationship of climate, which can limit its use for spatial impact assessment. Missing or erroneous data set can also lead to inappropriate outputs (Wilby et al., 2009). There are certain tools available to interpolate the parameters of weather generators over specific location. For instance, MarkSim GCM is a weather generating software and is based on existing GCM projections. It is set in Google Earth and can give daily data of future climatology over any point on the earth (CCAFS, 2011).

For areas, where adequate observed data are present for model calibration, a range of statistical downscaling models, from regressions to artificial neural network (ANN) and analogues, have been developed and applied. For example, Ghosh and Katkar (2012) downscaled the GCM future P_{cp} data for Assam and Meghalaya meteorologic subdivisions in India, using LR, ANN, and support vector machine (SVM). Tripathy et al. (2011) downscaled the future climate data for Indian Punjab by using environment visual image (ENVI) software to transform the global data to image format, then resizing it according to the baseline data boundary, and making the spatial resolution 1×1 degree to match the spatial resolution of baseline weather. Daily weather data on T_{max}, T_{min}, and P_{cp} have been created through addition/subtraction of monthly change (assumed same change for each day in the respective month) to the daily observed data (IPCC-TGICA, 2007). The monthly change factors for T_{max}, T_{min}, and P_{cp} have been computed from the GCM UKMO, HadCM3 (United Kingdom Meteorological Office, Hadley Climate Prediction Centre, Mitchell, Johns, Eagles, Ingram, & Davis, 1999), against the baseline averaged weather data of 1961–91.

2.5.2 Dynamic Downscaling

In dynamic downscaling, a limited-area model is derived from a parent GCM. It is often called a regional climate model (RCM) and is used to generate climate change scenarios at a higher spatial and temporal resolution of about 30–50 km and a time step size of 6 h for one particular period in the

present and one in the future (Mearns et al., 2003). Coarse-grid GCM simulation output is used for initial and lateral boundary conditions, known as
"one-way nesting approach" (Mearns et al., 2003), shown in Fig. 2.2. The
"nested" RCM approach was first applied in climate change studies in the
late 1980s by Dickinson, Errico, Giorgi, and Bates (1989). Even though
one-way nested approach, without any feedback from RCM to GCM, is
generally being used in many RCM simulations, the two-way nesting, with
feedback from RCM simulations back to the GCM, is a possible alternative
(Bowden, Otte, Nolte, & Otte, 2011; Chan et al., 2013; Foley, 2010;
Lorenz & Jacob, 2005). In RCMs, downscaled from GCMs, knowledge
of various climate processes controlling simulation of climate features such
as orographic P_{cp}, extreme climate events, and regional-scale climate anom-
alies or nonlinear effects associated with the El Niño-Southern Oscillation is
improved, especially in regions where land surface characteristics regulate
the regional distribution of climate variables (Wang et al., 2004).

The improvement in climate simulation at regional scales by RCM is due
to accounting for (i) mesoscale grid characteristics of existing topographic
features such as lakes, land-sea contrast, and land cover in homogeneity
and (ii) enhanced simulation of atmospheric circulations and climate at finer
spatial scales (IPCC-TGICA, 2003). The RCMs are being adopted worldwide as they offer many advantages such as (i) giving intense events that will
be smoothened in coarse resolution but may still miss the most extremes;
(ii) giving phenomenological values diurnal cycle; (iii) having more numerical stability and accuracy as these cover only a fraction of the globe and
require short-time steps; (iv) providing improvement in climate simulations,

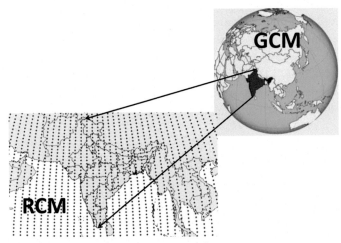

Fig. 2.2 Regional climate model nesting approach.

especially for P_{cp} that has high spatial variability; and (v) boundary conditions based on actual observations that provide information on fine-scale climate behavior besides isolating GCM error from the errors intrinsic to RCM. The RCMs have some limitations too: for example, (i) simulation is dependent upon the boundary conditions supplied from other source; (ii) climate needs parameterization for subgrid-scale processes, surface atmosphere coupling, and radiation transfer and cloud microphysics; (iii) only a limited number of scenario runs are available, and the "time slice" approach is used; (iv) the outputs of the dynamically downscaled RCM are dependent on the precision of the forcing GCM and its unfairness (Seaby et al., 2013); (v) the outputs of the RCM are liable to systematic errors and may involve a method of bias removal in addition to downscaling techniques for higher resolution; and (vi) climate information in grid-box size of an RCM is at higher resolution (>10 km) than local- or station-scale, which make such downscaling inappropriate for hydrologic and agricultural impact studies (Benestad, 2009).

The commonly employed RCMs in climate change studies are the Canadian regional climate model (CRCM); UK Met Office Hadley Centre's regional climate model version 3 (HadRM3); German regional climate model (REMO); Dutch regional atmospheric climate model (RACMO); US regional climate model version 3 (RegCM3); and German HIRHAM, which combines the dynamics of the high-resolution limited-area model (HIRLAM) and European Centre Hamburg model (ECHAM) (Trzaska & Schnarr, 2014). For Indian region, GCM data from the Hadley Centre's coupled model (HadCM3) were downscaled for providing a high-resolution RCM under the "Providing Regional Climate for Impact Studies" (PRECIS) project (Kumar et al., 2006, 2011). Nowadays, many tools with user-friendly interface are available to facilitate the downscaling of GCM data for wider utilization of available projections. Links to different portals for downscaling GCM outputs to regional/local scale have been compiled by Trzaska and Schnarr (2014). For example, the link for statistical downscaling model (SDSM) is http://co-public.lboro.ac.uk/cocwd/SDSM/, which comprises a software package for producing site-specific daily scenarios of climate variables and statistical parameters.

2.5.3 Choice of Downscaling Techniques

The use of downscaling techniques is dependent on the geographical location, type of predictors and predictands, climate data, and spatial and temporal resolutions. Most downscaling techniques are developed for a particular application, for example, biodiversity, agriculture, and water resources. Additionally, the suitability of a downscaling technique is region-/location-specific. That is why developing a statistical downscaling model is not preferred over RCM for a large number of sites as it is quite time-intensive and costly and requires very extensive observational data. The RCMs are preferred where purpose is to improve grid-scale climate modeling because these are able to resolve future on finer scales (covering a limited area) than those resolved by GCM, particularly related to improve resolution of the topography. However, due to the presence of intermodel inconsistency and model bias, it is proposed that an ensemble of RCM should be used instead of single-RCM approach (Teutschbein & Seibert, 2010). The coordinated RCM ensemble experiments within scientific projects such as Prediction of Regional scenarios and Uncertainties for Defining European Climate Change risks and Effects (PRUDENCE) (Christensen & Christensen, 2007), the ENSEMBLES project supported by the European Commission's Fifth and Sixth Framework Program, respectively, and North American Regional Climate Change Assessment Program (NARCCAP) (Mearns, 2009) have resulted in a more comprehensive range of RCM simulations using various RCMs, GCM forcings, and climate scenarios. In India, multimodel climate projections of surface temperature and monsoon P_{cp} for 1901–2098 were made available using the CMIP3 models by Kumar et al. (2010). Earlier, RCM projections from PRECIS were used to simulate the regional climate of India for the baseline (1961–90) and long-term climatology (2071–2100) for the SRES scenarios A2 and B2 (Kumar et al., 2006) and A1B scenario (Kumar et al., 2011). The most notable advantage of using the RCM is more realistic representation of spatial pattern of summer monsoon P_{cp}.

2.6 BIAS IN DOWNSCALED DATA AND ITS REMOVAL

The raw outputs of the climatic parameters from RCM models often suffer from systematic discrepancies between climate model simulations and observations, which may prevent their direct application for knowing the climate behavior and its local impacts. The discrepancies may arise from structural uncertainty caused by a finite number of variables. The discrepancy in modeled daily P_{cp} and temperature may afflict the monthly or annual time trends and magnitude. Classic biases may include the incidence of excessive wet events with low-intensity P_{cp} or inaccurate prediction of extreme temperatures in RCM simulations (Ines & Hansen, 2006). These biases are particularly distinct for P_{cp} than temperature (Andreasson, Bergstrom, Carlsson, Graham, & Lindstrom, 2004). Therefore, the projected raw data must be made bias-free by means of some corrections based on statistical corrective methods before use (Feddersen & Andersen, 2005). A number of statistical correction techniques from simple to advance for removing the bias in P_{cp} and temperature have been quoted in the literature (Boberg, Berg, Thejll, & Christensen, 2007; Teutschbein & Seibert, 2010). While using the corrective methods, the underlying assumptions are that there is no change in climate model biases with time and the corrections methods and their parameters are valid for longer area and stay steady with time, especially from baseline simulation to scenario generation (Deque, 2007; Hashino, Bradlley, & Schwartz, 2007).

2.6.1 Correction of Modeled Data With Reference to the Observed

The main two strategies used in calibration of climate projections are bias correction and change factor (Ho, Stephenson, Collins, Ferro, & Brown, 2012). Bias correction is the procedure in which climate model outputs are scaled to account for systematic errors in the climate models caused by flawed conceptualization, grid selection, and spatial averaging within each grid cell. There are seven methods, ranging from simple (precipitation threshold, scaling approach, linear transformation, and power transformation) to advance (distribution transfer, precipitation model, and empirical correction), by which P_{cp} and temperature can be bias-corrected. These methods are commonly used to bring the modeled data near to the observed in regard to magnitude and time trends. However, the improvements to the statistical properties of the data are limited to the specific timescale of the fluctuations and the site. The worldwide researchers commonly apply simple methods like difference and statistical bias correction (SBC) for correcting modeled data, which are given below.

Difference method: In this method, averaged across past 20–30 years, daily difference of observed and modeled values (ΔP_{cp}) of a climate parameter, P_{cp}, is taken for each Julian day (365 days). The ΔP_{cp} is considered as daily correction factor, which is added to the modeled uncorrected value (P_{cp} model uncor.) to correct it (P_{cp} model cor.) so that the values approach the observed ones (Eq. 2.1):

$$P_{cp}\,\text{model cor.} = P_{cp}\text{model uncor.} + \left(\Delta P_{cp}\right) \qquad (2.1)$$

Modified difference method: The modified difference method is similar to the difference method. However, some statistical parameters like mean (μ) and standard deviation (σ) are added aiming at shifting and scaling to adjust the μ and σ and improve the correction function. For P_{cp}, the relation used is Eq. (2.2).

$$P_{cp}\,\text{model cor.} = \left(P_{cp}\,\text{modeled unncor.} + \left(\sigma P_{cp}\right)\right) x \left(\frac{\mu P_{cp}\,\text{observed}}{\mu P_{cp}\,\text{modeled}}\right)$$

(2.2)

For temperature (T) correction, the relation is Eq. (2.3) (Leander & Buishand, 2007).

$$\mu T_{cor.} = \mu T_{obs.} + \left(\sigma T_{obs.}/\sigma T_{model}\right)$$
$$\times \left(T_{uncor.} - \mu T_{obs}\right) + \left(\mu T_{obs.} - \mu T_{model}\right)$$

(2.3)

Statistical bias correction: It is a statistical procedure, which involves fitting of the probability density function (PDF) of model data to that of the observations. It is used to rectify the cumulative distribution function (CDF) of the future modeled data corresponding to the observed data in climate change studies. In SBC method, if x denotes the considered variable (temperature or P_{cp}) and $F(x)$ denotes the CDF of x (Eq. 2.4), then transformation that changes the particular daily value of RCM model run for control period (x_{mod}) to corrected (bias-corrected) value of it ($x_{mod\ cor}$) at a specified probability is

$$x_{mod\ cor} = F_{obs}^{-1}\,F_{mod}\left(x_{mod}\right)$$

(2.4)

Here, F_{mod} is CDF of x for RCM model, and F_{obs}^{-1} is the inverse of observed CDF of x. The procedure sometimes involves splitting the series into its different timescales, known as cascade bias correction (Haerter, Hagemann, Moseley, & Piani, 2010) to account for the minor fluctuations of temperature or P_{cp} in some months of a year, which may occur due to the systematic seasonal dependence of statistical expectation value within the month. Such corrections have already been applied separately for P_{cp} (Piani, Haerter, & Coppola, 2010) and temperature (Sennikovs & Bethers, 2009) data. For P_{cp}, intensity-based correction gives better results (Piani, Weedon, et al., 2010) than total and frequency. While selecting bias correction method, it is very important that bias correction should improve not only the statistical parameters but also the time trends. In case, a number of stations are to be considered, mean bias ratio (Eq. 2.5) is preferred.

$$\text{Mean bias ratio} = \text{RCM model value of } a \text{ variable/observed value} \quad (2.5)$$

Jalota, Kaur, Kaur, and Vashisht (2013), Kaur (2013), and Kaur et al. (2015) corrected the modeled data derived from PRECIS model by employing such methods and projected crop yield and water balance in rice-wheat cropping system for mid century (MC, 2021–50) and end century (EC, 2071–98).

2.7 CLIMATE CHANGE TRENDS AND PROJECTIONS

As discussed in Chapter 1, atmospheric concentrations of GHGs have increased markedly at global level. It was Fourier (1827), who first postulated that increase in GHGs in the atmosphere could lead to an increase in surface temperature. Over the last 100 years, there is an accepted global change. The global warming was focused into the period of 1910–45. The magnitude and trend of the observed climate change varied at global, regional, and local scale. As per fourth assessment report of IPCC (2007), 11 of the 12 years (1995–2006) rank among the 12 warmest years in the instrumental record of global surface temperature since 1850. The climate system was warmed by 0.74°C from 1906 to 2005 (Ghude, Jain, & Arya, 2009; Sathaye, Shukla, & Ravindranath, 2006). The warming is not uniform everywhere and every time. For example, surface air over lands warms more and makes the hydrologic cycle more intense due to holding more moisture than that over the ocean. Warming rate is more in arid zones than humid (IPPC, 2007). Over the last 50 years, some extreme weather events have changed. Cold days, cold nights, and frosts have become less frequent over most land area, while hot days, hot nights, and heat waves are becoming frequent. Averaged Antarctic temperatures have increased at almost twice the global average rate in the past 100 years. The warming has been much more pronounced in the recent decades, with more increase in night temperatures as compared with that of daytime. In India, a considerable warming of about 0.3–0.6°C in the average annual surface air temperature since the 1860s has been observed (Hingane, Kumar, & Ramana Murty, 1985; Pant, 2003), which is comparable with the global-mean trend of 0.5°C in the last 100 years (Sathaye et al., 2006).

The snow cover is also believed to be gradually decreasing (De la Mare, 2009). In some places, climatic extremes such as droughts, floods, timing of P_{cp}, and snowmelt have also increased. Frequency of heavy P_{cp} events has increased over most area. Even though the representative P_{cp} series across various locations in India is trendless, it exhibits a highly variable pattern for more than a century with a distinctive epochal nature of variability. But at certain subregions, statistically significant increasing and decreasing trends were present (Pant & Kumar, 1997; Sontakke, 1990). The number of dry days may increase by 15 days in the western and central regions, whereas north and northeastern areas may experience 5–10 more days of P_{cp} annually indicating that dry areas will get drier and wet areas wetter (Pant, 2003; Pant & Kumar, 1997).

Sea level, consistent with warming, has increased at a broad range of sites worldwide since 1995. Global sea level rose by an average rate of 3.1 m year^{-1} from 1993 to 2003. Sea-level rise is contributed by thermal expansion of the oceans (57%), decrease in glaciers and ice caps (28%), and losses from polar ice sheets (15%). For Indian coastal regions, the sea level has risen by 100–200 mm with regional variations (Gosain, Rao, & Debajit Basuray, 2006; Unnikrishnan, Kumar, Fernandes, Michael, & Patwardhan, 2006).

Future climate change is unfolded by projections, prediction, and forecast. Prediction is any depiction of the future and pathways leading to it; prediction is when projection is branded; and forecast is obtained using deterministic models. Predictions for future climate change are made from the data simulated by the climate models, namely, GCMs and RCMs based on the gas emission scenarios explained in SRES and RCPs. Based on SRES scenario, the best estimates of temperature increase at the end of the 21st century compared with year 2000 were lowest (1.2°C) in B1 and highest (3.4°C) in A1F scenario (IPCC, 2007). For the period 2016–35 relative to 1986–2005, the global average surface temperature change will increase in the range 0.3–0.7°C (medium confidence) and is similar for the four RCPs (IPCC, 2014) under the assumption that there will be no volcanic eruptions or any unexpected changes in total solar irradiance. After MC, the enormity of the projected climate change is significantly dependent on the choice of emission scenario. Under RCP 2.6, the global-mean surface temperature is likely to increase by 0.3–1.7°C at the end of the twenty-first century (2081–2100) relative to 1986–2005. The increase in the global-mean surface temperature is likely to be 1.1–2.6°C under RCP 4.5, 1.4–3.1°C under RCP 6.0, and 2.6–4.8°C under RCP 8.5. The Arctic region will warm more rapidly than the global mean. Changes in P_{cp} will not be the same. Under the RCP 8.5 scenario, the high latitudes and the equatorial Pacific are expected to experience a rise in annual average P_{cp}, whereas P_{cp} is likely to decrease in many midlatitude and subtropical dry regions and increase in many midlatitude wet regions. Extreme P_{cp} events likely become more intense and frequent over most of the midlatitude landmasses and over wet tropical regions. It is expected that heat waves shall be more frequent and will last longer, and the intensity and frequency of extreme P_{cp} events will increase in many regions (IPCC, 2014).

For Indian region, Chaturvedi, Joshi, Jayaraman, Bala, and Ravindranath (2012) projected mean warming 2°C in RCP 2.6 and 4.8°C in RCP 8.5 from the 1880s to the 2080s through CMIP5 model ensemble-based projections. All India P_{cp} is projected to increase by 6%, 10%, 9%, and 14% under the scenarios of RCP 2.6, RCP 4.5, RCP 6.0, and RCP 8.5, respectively, by the 2080s relative to the 1961–90 bases. However, much larger variability will be perceived in the spatial distribution of P_{cp}. But Kumar et al. (2006) projected a temperature rise of 2.9°C and 4.1°C for India under the B2 and A2 scenarios of SRES, respectively, in the 2080s relative to the 1970s. Over the same period, projected warming for the A1B scenario is 3.5–4.3°C (Kumar et al., 2011). Though variation among models, scenarios, and agroclimatic zones exist, trends show an increase in temperature and P_{cp}. For example, Jalota and Vashisht (2016) projected that, averaged across three agroclimatic zones (hot humid, hot semiarid, and hot typical arid), three GCMs (HadCM3, CSIRO-Mk2, and CCCMA-CGCM2), and two scenarios (A2 and B2), T_{max} would be more by 1.1°C, 2.2°C, and 3.6°C; T_{min} by 1.7°C, 3.0°C, and 4.2°C; and P_{cp} by 38, 49, and 57 mm in 2020, 2050, and 2080, respectively, compared with 1998–2008 in Punjab State of India. With RCM-PRECIS for A1B scenario, the projected T_{max}, T_{min}, and P_{cp} would increase in the future at the rate of 0.052°C, 0.053°C, and 4.83 mm year^{-1}, respectively (Fig. 2.3A–C). However, the change varied with the location and time segment (Jalota, Vashisht, Kaur, Kaur, & Kaur, 2014). For example, T_{max} at Amritsar (Fig. 2.3D) and T_{min} at Patiala (Fig. 2.3E) than other locations in PTS (1989–2010) would continue in MC (2021–50) and EC (2071–98). Higher P_{cp} at Jalandhar in PTS would shift to Patiala in both MC and EC (Fig. 2.3F).

It is not only the annual but also the seasonal and monthly trends that would change (Table 2.3 and Fig. 2.4).

It is not only the annual but also the seasonal (Table 2.3) and monthly (Fig. 2.4) trends that would change. For example, at Ludhiana location,

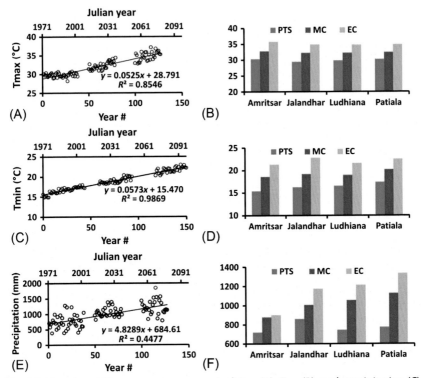

Fig. 2.3 Time trend (5-year moving average) of T_{max} (A), T_{min} (B), and precipitation (C); variation in different time slices at four locations in T_{max} (D), T_{min} (E), and precipitation (F).

Table 2.3 Seasonal Climate Change in Different Time Slices

Time Slice	Season	T_{max} (°C)	T_{min} (°C)	Precipitation (mm)
Present time slice	Kharif	34.0±0.7	24.2±0.9	574±88
	Rabi	26.9±0.1	12.0±0.9	240±40
Mid century	Kharif	37.7±1.3	27.7±1.1	739±119
	Rabi	29.3±1.7	14.2±0.7	322±104
End century	Kharif	40.7±1.7	31.2±1.3	878±228
	Rabi	31.7±1.4	16.3±0.5	331±100

Present time slice, mid century, and end century represent periods of 1989–2008, 2021–50, and 2071–98, respectively.

Reproduced from Kaur, S. (2013). *Modelling the impact of climate change on groundwater resources in central Punjab* (PhD dissertation). Ludhiana, India: Punjab Agricultural University.

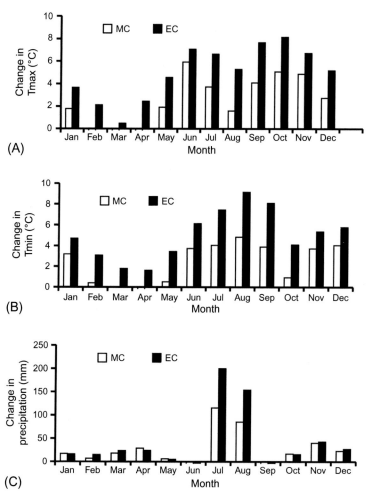

Fig. 2.4 Monthly change in maximum temperature (A), minimum temperature (B), and precipitation (C) trends in mid century (MC) and end century (EC) compared with present time slice (PTS).

during wet and hot season (locally named as *Kharif*), T_{max} would elevate by 3.7°C in MC and 6.7°C in EC compared with PTS. The corresponding elevation during dry and cold season (locally named as *Rabi*) would be by 2.4°C in MC and 4.8°C in EC. T_{min} during *Kharif* season would elevate by 3.5°C in MC and 7.0°C in EC and during *Rabi* season by 2.2°C in MC and 4.3°C in EC. Precipitation would increase by 165 mm in MC and 304 mm in EC during *Kharif* season and 82 mm in MC and 91 mm in

EC during *Rabi* season. In MC, the change in monthly T_{max} would be positive in all months except in February, March, and April.

Highest positive change would be of 5.8°C in the month of June and negative change of 1.3°C in the month of March. In EC, the change in T_{max} would be positive in all the months, and the maximum positive would be 8.1°C in the month of October (Fig. 2.4A).

On monthly basis, in MC, there would be positive change in T_{min} in 11 months with highest of 4.8°C in the month of August and negative in 1 month of April by 0.4°C. In EC, the change would be positive in all the months with highest of 9.1°C in the month of August (Fig. 2.4B). Monthly trends showed that change in P_{cp} would be positive in all the months in MC and EC compared with that in the PTS, except in the months of June and September (Fig. 2.4C). The highest positive change in P_{cp} would be in the month of July, which was computed as 104 mm in MC and 193 mm in EC.

All these trends predict that in the future, increase in T_{max}, T_{min}, and P_{cp} is location-specific; therefore, adaptation technologies and mitigation strategies have to be developed according to the scale of climate change.

EXERCISES

1. Choose the best answer
    i. Climate projections from AOGCMs are limited for regional applications due to _____.
        a. coarse spatial resolution
        b. 30-year-time steps
        c. biases in the model
        d. the lack of emissions scenario data
    ii. Using a simulation ensemble provides more information about trends and spread than using a single projection.
        a. True
        b. False
    iii. Bias correction describes a process for removing _____ in the AOGCM data.
        a. random errors
        b. major trends with time in the future
        c. limitations caused by coarse spatial resolution
        d. systematic errors

    iv. Climate projections for future time periods of decades and centuries are mainly _____ problem.

      a. a spatial resolution

      b. an initial value

      c. an emission scenario

      d. a boundary value

    v. The differences between weather predictions and climate projections can be described as follows:

      a. Weather predictions are for specific events for time periods of hours and days.

      b. Climate projections are statistical descriptions for future periods only, for time periods including years and decades.

      c. Climate projections are statistical descriptions, past or future, for time periods including years and decades.

      d. Weather predictions lack probability.

    vi. A possible advantage of using an ensemble of projections rather than one single projection to represent a climate change scenario is that an ensemble _____.

      a. expands the range of the possible changes

      b. does not rely on bias-corrected simulation data

      c. helps us focus more on the dominant trend in the data

      d. can be used with the transient method

    vii. The delta method for generating climate-adjusted weather inputs differs from the ensemble-informed delta method in that _____.

      a. delta only uses one projection for each climate change scenario

      b. delta tends to smooth out the variability in the future simulations

      c. delta uses a mean of multiple projections for each climate change scenario

      d. delta change factors tend to vary more significantly from month to month

    viii. _____pathways of possible future changes in greenhouse gas concentrations.

    ix. _____ set of possible mean characteristics of a future climate, for example, hotter and wetter.

    x. _____widely used emission scenarios based on IPCC 2000 report.

      Answers (i, a; ii, a; iii, d; iv, d; v, a and c; vi, c; vii, a and d; viii, Representative concentration; ix, Climate change scenario; and x, Emission scenario).

2. Develop correction functions, and calculate the bias-corrected values of maximum and minimum temperatures for January 1, 2040–50 from the observed and modeled data for the years from 1971 to 1990 (Table 2.4) and modeled uncorrected data for the years from 2040 to 2050 (Tables 2.5 and 2.6).

Table 2.4 Modeled and Observed Temperature Data for the Years 1971–90

Date	Modeled $^aT_{max}$ (°C)	Modeled $^aT_{min}$ (°C)	Observed T_{max} (°C)	Observed T_{min} (°C)
January 1, 1971	16.6	1.2	20.0	1.6
January 1, 1972	13.5	1.3	20.0	3.6
January 1, 1973	15.9	3.3	18.5	1.2
January 1, 1974	15.6	2.1	17.8	1.2
January 1, 1975	15.5	−2.0	14.2	9.1
January 1, 1976	14.2	−1.9	19.7	7.0
January 1, 1977	23.4	9.6	18.0	1.5
January 1, 1978	15.1	2.2	16.7	3.8
January 1, 1979	13.8	−1.7	20.6	2.9
January 1, 1980	13.9	−0.1	14.1	10.6
January 1, 1981	20.0	0.7	15.2	5.0
January 1, 1982	13.7	2.4	20.4	5.4
January 1, 1983	17.4	2.1	15.2	6
January 1, 1984	21.0	1.7	19.2	3.2
January 1, 1985	16.9	0.8	11.8	7.5
January 1, 1986	20.1	5.2	17.6	6.0
January 1, 1987	14.0	1.2	19.2	2.8
January 1, 1988	18.6	3.1	19.0	6.5
January 1, 1989	21.1	3.9	20.0	4.8
January 1, 1990	12.5	0.4	11.4	3.8

$^a T_{max}$ is maximum temperature, and T_{min} is minimum temperature.

Table 2.5 Uncorrected Modeled Data

Years	Maximum Temp (°C)	Minimum Temp (°C)
January 1, 2040	18.1	5.5
January 1, 2041	14.3	2.0
January 1, 2042	21.8	3.7
January 1, 2043	16.7	5.4
January 1, 2044	21.2	15.8
January 1, 2045	28.3	16.4
January 1, 2046	20.1	4.1
January 1, 2047	16.3	8.4
January 1, 2048	19.2	4.4
January 1, 2049	21.5	7.6
January 1, 2050	22.0	3.2

Answers (Correction factors for maximum and minimum temperatures are −0.79 and 2.92. Corrected modeled data are in Table 2.6).

Table 2.6 Corrected Modeled Data

Year	Difference Method		Modified Difference Method	
	T_{max} (°C)	T_{min} (°C)	T_{max} (°C)	T_{min} (°C)
January 1, 2040	17.3	2.6	18.8	8.4
January 1, 2041	13.5	−0.9	15.4	5.0
January 1, 2042	21.0	0.8	22.2	6.6
January 1, 2043	15.9	2.5	17.6	8.3
January 1, 2044	20.4	12.9	21.6	18.3
January 1, 2045	27.5	13.5	28.1	18.9
January 1, 2046	19.3	1.2	20.6	7.0
January 1, 2047	15.5	5.5	17.2	11.2
January 1, 2048	18.4	1.5	19.8	7.3
January 1, 2049	20.7	4.7	21.9	10.4
January 1, 2050	21.2	0.3	22.4	6.2

REFERENCES

Andreasson, J., Bergstrom, S., Carlsson, B., Graham, L. P., & Lindstrom, G. (2004). Hydrological change—Climate change impact simulation for Sweden. *Ambio, 33*, 228–234.

Benestad, R. E. (2008). *Empirical-statistical downscaling* [pp. 228]. Singapore: World Scientific Publishing Co. Pvt. Ltd.

Benestad, R. E. (2009). Downscaling precipitation extremes. *Theoretical and Applied Climatology, 100*, 1–21.

Bjornaes, C. (2013). *A guide to representative concentration pathways* (pp. 1–5). Norway: CICERO, Center for International Climate and Environmental Research.

Boberg, F., Berg, P., Thejll, P., & Christensen, J. H. (2007). *Analysis of temporal changes in precipitation intensities using PRUDENCE data.* Danish Climate Centre Report 7, no.3, Copenhagen.

Bowden, J. H., Otte, T. L., Nolte, C. G., & Otte, M. J. (2011). Examining interior grid nudging techniques using two-way nesting in the wrf model for regional climate modeling. *Journal of Climate, 25*, 2805–2823.

CCAFS (2011). In *MarkSim GCM: generating plausible weather data for future climates. Using Climate Scenarios and Analogues for Designing Adaptation Strategies in Agriculture.* Kathmandu, Nepal.

Chan, S. C., Kendon, E. J., Fowler, H. J., Blenkinsop, S., Ferro, C. A. T., & Stephenson, D. B. (2013). Does increasing the spatial resolution of a regional climate model improve the simulated daily precipitation? *Climate Dynamics, 41*, 1475–1495.

Chaturvedi, R. K., Joshi, J., Jayaraman, M., Bala, G., & Ravindranath, N. H. (2012). Multimodel climate change projections for India under representative concentration pathways. *Current Science, 103*, 791–802.

Christensen, J. H., & Christensen, O. B. (2007). A summary of the PRUDENCE model projections of changes in European climate by the end of this century. *Climate Change, 81*, 7–30.

De la Mare, W. K. (2009). Changes in Antarctic sea-ice extent from direct historical observations and whaling records. *Climatic Change, 92*, 461–493.

Deque, M. (2007). Frequency of precipitation and temperature extremes over France in an anthropogenic scenario; model results and statistical correction according to observed values. *Global and Planetary Change, 57*, 16–26.

Dickinson, R. E., Errico, R. M., Giorgi, F., & Bates, G. T. (1989). A regional climate mod-elfor the western United States. *Climatic Change, 15*, 383–422.

Feddersen, H., & Andersen, U. (2005). A method for statistical downscaling of seasonal ensemble predictions. *Tellus, 57A*, 398–408.

Flato, G. (2011). Earth system models: an overview. *Wiley Interdisciplinary Reviews Climate Change, 2*, 783–800.

Foley, A. M. (2010). Uncertainty in regional climate modelling: A review. *Progress in Physical Geography, 34*, 647–670.

Fourier, J. B. J. (1827). *Memorandum on temperatures of the terrestrial globe and planetaryspace:* (Vol. 7, pp. 569–604).

Ghosh, S., & Katkar, S. (2012). Modeling uncertainty resulting from multiple downscaling methods in assessing hydrological impacts of climate change. *Water Resources Management, 26*, 3559–3579.

Ghude, S. D., Jain, S. L., & Arya, B. (2009). Temporal evolution of measured climate forcing agents at South Pole, Antartica. *Current Science, 96*, 49–57.

Gosain, A. K., Rao, S., & Debajit Basuray, D. (2006). Climate change impact assessment on hydrology of Indian River basins. *Current Science, 90*, 346–353.

Haerter, J. O., Hagemann, S., Moseley, C., & Piani, C. (2010). Climate model bias correction and the role of timescale. *Hydrology and Earth System Sciences Discussion, 7*, 7863–7898.

Hashino, T., Bradlley, A. A., & Schwartz, S. S. (2007). Evaluation of bias-correction methods ensemble streamflow water forecast. *International Journal of Earth Sciences and Engineering, 11*, 939–950.

Hingane, L. S., Kumar, R. K., & Ramana Murty, B. V. (1985). Long-term trends of surface air temperature in India. *Journal of Climatology, 5*, 521–528.

Ho, C. K., Stephenson, D. B., Collins, M., Ferro, C. A., & Brown, S. J. (2012). Calibration strategies: A source of additional uncertainty in climate change projections. *Bulletin of the American Meteorological Society, 93*, 21–26.

Ines, A. V. M., & Hansen, J. W. (2006). Bias correction of daily GCM rainfall and crop simulation studies. *Agricultural and Forest Meteorology, 138*, 44–53.

IPCC. (2007). *Climate change: Climate change impacts, adaptation and vulnerability.* Working Group II contribution to Intergovernmental panel on climate change Fourth Assessment Report. Summary for policy makers, (pp. 23).

IPCC. (2014). In Core writing Team, R. K. Pachauri, & L. A. Meyers (Eds.), *Climate change 2014 synthesis report. Contributions of Working Group I, II, III to the Fifth Assessment Report of Inter Governmental Panel on Climate Change.* Geneva Switzerland: IPCC. (pp. 151) in IPCC. AR5 Synthesis Report website.

IPCC-SRES. (2000). In N. Nakićenovic & R. Swart (Eds.), *Special report on emissions scenarios: A special report of Working Group III of the Intergovernmental Panel on Climate Change.* Cambridge University Press. ISBN 0-521-80081-1, 978-0-521-80081-5 (pb: 0-521-80493-0, 978-0521-80493-6).

IPCC-TGICA. (2003). *Guidelines for use of climate scenarios developed from regional climate model experiments. Prepared by T.R. Carter on behalf of the IPCC, Task Group on Data and Scenario Support for Impact and Climate Assessment,* (pp. 66). Retrieved from www.ipcc-data.org/guidelines.

IPCC-TGICA. (2007). General guidelines on the use of scenario data for climate impact and adaptation assessment. Version 2. Prepared by T.R. Carter on behalf of the Intergovernmental Panel on Climate Change, Task Group on Data and Scenario Support for Impact and Climate Assessment, (pp. 66).

Jalota, S. K., Kaur, H., Kaur, S., & Vashisht, B. B. (2013). Impact of climate change scenarios on yield, water and nitrogen-balance and-use efficiency of rice-wheat cropping system. *Agricultural Water Management, 116*, 29–38.

Jalota, S. K., & Vashisht, B. B. (2016). General circulation models driven climate change in different agro-climatic zones of Indian Punjab: ensuing crop productivity. *Journal of Agrometeorology*, *18*, 48–56.

Jalota, S. K., Vashisht, B. B., Kaur, H., Kaur, S., & Kaur, P. (2014). Location specific climate change scenario and its impact on rice and wheat in Central Indian Punjab. *Agricultural Systems*, *131*, 77–86.

Kaur, S. (2013). *Modelling the impact of climate change on groundwater resources in central Punjab* [PhD dissertation]. Ludhiana, India: Punjab Agricultural University.

Kaur, S., Jalota, S. K., Kaur, H., Vashisht, B. B., Jalota, U. R., & Lubana, P. P. S. (2015). Evaluation of statistical corrective methods to minimize bias at different time scales in a regional climate model driven data. *Journal of Agrometeorology*, *17*, 29–35.

Kumar, K. K., Kamala, K., Rajagopalan, B., Hoerling, M. P., Eischeid, J. K., Patwardhan, S. K., et al. (2010). The once and future pulse of Indian monsoonal climate. *Climate Dynamics*, *36*, 2159–2170.

Kumar, K. K., Patwardhan, S. K., Kulkarni, A., Kamala, K., Rao, K., & Jones, R. (2011). Simulated projections for summer monsoon climate over India by a high-resolution regional climate model (PRECIS). *Current Science*, *101*, 312–326.

Kumar, R. K., Sahai, A. K., Kumar, K. K., Patwardhan, S. K., Mishra, P. K., Revadekar, J. V., et al. (2006). High-resolution climate change scenarios for India for the 21st century. *Current Science*, *90*, 334–345.

Leander, R., & Buishand, T. A. (2007). Resampling of regional climate model output for the simulation of extreme river slopes. *Journal of Hydrology*, *332*, 487–496.

Lorenz, P., & Jacob, D. (2005). Influence of regional scale information on the global circulation: A two-way nesting climate simulation. *Geophysical Research Letters*, *32*, L18706.

Mearns, L. O., Giorgi, F., Whetton, P., Pabon, D., Hulme, M., & Lal, M. (2003). *Guidelines for use of climate scenarios developed from Regional Climate Model experiments.* Data Distribution Centre of the Intergovernmental Panel on Climate Change.

Mearns, L., & NARCCAP Team (2009). The North American Regional Climate Change Assessment Program (NARCCAP): Overview of Phase II results. *IOP Conference Series: Earth and Environmental Science*, *6*, 022007.

Mitchell, J. F., Johns, T. C., Eagles, M., Ingram, W. J., & Davis, R. A. (1999). Towards the construction of climate change scenarios. *Climatic Change*, *41*, 547–581.

Pant, G. B. (2003). Long-term climate variability and change over monsoon Asia. *Journal of Indian Geophysical Union*, *7*, 125–134.

Pant, G. B., & Kumar, R. K. (1997). *Climates of South Asia* [pp. 320]. Chichester: John Wiley & Sons.

Piani, C., Haerter, J. O., & Coppola, E. (2010). Statistical bias correction for daily precipitation in regional climate models over Europe. *Theoretical and Applied Climatology*, *99*, 187–192.

Piani, C., Weedon, G. P., Best, M., Gomes, S. M., Viterbo, P., Hagemann, S., et al. (2010). Statistical bias correction of global simulated daily precipitation and temperature for the application of hydrological models. *Journal of Hydrology*, *395*, 199–215.

Sathaye, J., Shukla, P. R., & Ravindranath, N. H. (2006). Climate change, sustainable development and India: Global and national concerns. *Current Science*, *90*, 314–325.

Seaby, L. P., Refsgaard, J. C., Sonnenborg, T. O., Stisen, S., Christensen, J. H., & Jensen, K. H. (2013). Assessment of robustness and significance of climate change signals for an ensemble of distribution-based scaled climate projections. *Journal of Hydrology*, *486*, 479–493.

Sennikovs, J., & Bethers, U. (2009). R. S. Anderssen, R. D. Braddock, & L. T. H. Newham (Eds.), *Statistical downscaling method of regional climate model results for hydrological modelling* 18th World IMACS Congress and MODSIM09 International Congress on Modelling and Simulation, *Cairns, Australia. Modelling and Simulation Society of Australia and New Zealand and International Association for Mathematics and Computers in Simulation* (pp. 3962–3968).

Sontakke, N. A. (1990). Indian summer monsoon rainfall variability during the longest instrumental period 1813–1988. M.Sc Thesis Pune, India: University of Poona.

Team, C. W., Knutti, R., Abramowitz, G., Collins, M., Eyring, V., Gleckler, P. J., et al. (2010). T. F. Stocker, D. Qin, G. K. Plattner, M. Tignor, & P. M. Midgley (Eds.), *IPCC Expert Meeting on Assessing and Combining Multi Model Climate Projections Meeting Report of the Intergovernmental Panel on Climate Change Switzerland: University of Bern.*

Teutschbein, C., & Seibert, J. (2010). Regional climate models for hydrological impact studies at the catchment scale: A review of recent modeling strategies. *Geography Compass, 4,* 834–860.

Tripathy, R., Ray, S. S., Kaur, H., Jalota, S. K., Bal, S. K., & Panigrahy, S. (2011). *Understanding spatial variability of cropping system response to climate change in Punjab state of India using remote sensing data and simulation model ISPRS archives XXXVIII–8/W20. Workshop Proceedings: Earth Observations for Terrestrial Ecosystem* (pp. 29–33).

Trzaska, S., & Schnarr, E. (2014). *A review of downscaling methods for climate change projections* United States Agency for International Development by Tetra Tech ARD (pp. 1–42).

Unnikrishnan, A. S., Kumar, K. R., Fernandes, S. E., Michael, G. S., & Patwardhan, S. K. (2006). Sea level changes along the Indian coast: Observations and projections. *Current Science, 90,* 314–325.

Wang, Y. L. R., Leung, L. R., McGregor, J. L., Lee, D. K., Wang, W. C., Ding, Y., et al. (2004). Regional climate modeling: Progress, challenges and prospects. *Journal of the Meteorological Society of Japan, 82,* 1599–1628.

Weyant, J., Azar, C., Kainuma, M., Kejun, J., Nakicenovic, N., Shukla, P.R., et al. (2009). Report of 2.6 versus 2.9 Watts/m^2 RCPP evaluation panel. Technical Report, Intergovernmental Panel on Climate Change.

Wilby, R. L., Troni, J., Biot, Y., Tedd, L., Hewitson, B. C., Smith, D. M., et al. (2009). A review of climate risk information for adaptation and development planning. *International Journal of Climatology, 29,* 1193–1215.

Zorita, E., & Von Storch, H. (1999). The analog method as a simple statistical downscaling technique: Comparison with more complicated methods. *Journal of Climate, 12,* 2474–2489.

FURTHER READING

http://www.ipcc-data.org/guidelines/pages/gcm_guide.html.

Kaur, S., Jalota, S. K., & Aggarwal, R. (2014). Impact of climate change scenario on yield and irrigation water requirements of rice-wheat cropping system in central Punjab. *International Journal of Earth Sciences and Engineering, 7,* 296–302.

Miao, C., Duan, Q., Sun, Q., Huang, Y., Kong, D., Yang, T., et al. (2014). Assessment of CMIP5 climate models and projected temperature changes over Northern Eurasia. *Environmental Research Letters, 9,* 055007.

Van Vuuren, D. P., Edmonds, J., Kainuma, M., Riahi, K., Thomson, A., Hibbard, K., et al. (2011). The representative concentration pathways: An overview. *Climatic Change, 109,* 5–31.

> CHAPTER THREE

Climate Change Impact on Crop Productivity and Field Water Balance

Contents

> ## 3.1 INTRODUCTION

The climatic parameters affecting field crop productivity are maximum temperature (T_{max}), minimum temperature (T_{min}), carbon dioxide (CO_2), precipitation (P_{cp}), solar radiation (SR), and incidence of extreme

Understanding Climate Change Impacts on Crop Productivity and Water Balance
https://doi.org/10.1016/B978-0-12-809520-1.00003-3
87

events. It is evident from a number of reports that CO_2, T_{max}, T_{min}, and P_{cp} are going to change in the future, which is likely to influence the plants' life by changing physiological processes, soil environment, and field water-balance components. Rising CO_2 concentration in the atmosphere has an effect on soil carbon, population of beneficial soil microbes, soil structure, and nutrient acquisition/availability. In plants, elevated CO_2 affects photosynthesis, respiration, evapotranspiration, water–use efficiency (WUE), phenological development, yield attributes, and ultimately crop yield. Unlike CO_2, increased temperature influences the availability of carbon (C) and nitrogen (N) in soil. It is well known that each crop species has an optimum temperature range at each growth stage so that normal physiological processes go on for maximum productivity. Any gradual or sudden change in weather parameters, especially T_{max} and/or T_{min} from apposite at the most vulnerable crop growth stage and the number of days having greater or lesser than optimum temperatures, may adversely affect the crop yield and warrants defining the weather descriptors for different crops and locations. Increase in temperature alters the soil water status by increasing evaporation rate from soil and transpiration rate from plants, which causes soil moisture stress, increased irrigation water requirement, and hazard of salt accumulation in the soil. On the contrary, in some low-temperature regions, an increase in temperature may have positive effect on crop yield. Precipitation affects the plants through water retained in the soil, which directly affects the plant water status and nutrient availability. Extreme P_{cp} events can cause runoff and water erosion depending upon soil type and land use.

Despite crop yields, elevated temperature, CO_2, and P_{cp} affect the quality of the marketable produce too. All these effects of climate change on soil environment, plant development, and its yield are modified by moisture regime and N level in the soil, photosynthetic pathway (C3, C4, and CAM) a crop follows, and N-fixing capability of the crop. Besides change in climatic parameters, their variability also affects crop productivity.

The plant response to change in individual climate parameter has been studied in different regions of the world experimentally, mostly by free-air CO_2 enrichment (FACE) experiments, open-top chamber (OTC), temperature-gradient tunnel (TGT), etc. In the FACE experiments, concentration of CO_2 in a specified area is raised, and response on plant growth is measured. FACE experiments have a preference over studies conducted in laboratories as these allow studying the effect of elevated CO_2 on crops and plants, which are grown under field conditions keeping up the plant competition. Open-top chambers (OTC) are installed in the field (open plots) and are of different designs (Tiwari & Agrawal, 2006). Each of the chambers is attached to high-speed blowers with nonfiltered air in nonfiltered chambers and with filtered air through activated charcoal in filtered chambers. In the chambers, microclimatic measurements of temperature, relative humidity, CO_2, and light intensity are made. Temperature-gradient tunnel (TGT) is a clear plastic-coated tent in which different temperature above the ambient temperature using solar heating during the day and electric heating at night and enriched CO_2 are maintained. In case the experimentation is not time- and cost-effective to study the integrated effect of climate variables and their interactions among themselves and with soil environment (water and N), simulation approach is used through crop models (CropSyst, DSSAT, APSIM, EPIC, etc.). Crop simulation models are the mathematical expressions, which compute the impact of climate change employing relationships developed from the experimental data generated in the preceding studies. In the models, major input variables are daily weather data (T_{max}, T_{min}, P_{cp}, relative humidity (RH), SR, wind speed (WS), and atmospheric CO_2 concentration), soil profile data (sand, silt, clay, organic carbon, nitrate-nitrogen (NO_3—N), ammoniacal nitrogen (NH_4—N), and soil moisture content), plant characteristics (photosynthetic pathway and morphological, phenological sensitivity to heat and water stresses), management interventions (tillage, planting, irrigation, fertilizer, etc.). The crop models are able to generate information on daily biomass, daily water, and N balance components, crop phenology, and yield as influenced by climate variability and change, soil type, management interventions, and their interactions, which otherwise are difficult to ascertain from field experimentation.

In this chapter, fundamentals of direct and interactive effects of climate variables (CO_2, temperature, and P_{cp}) in the changing climate scenario on soil environment (labile C pools, microbial population and diversity, nutrient availability, and soil water evaporation) and processes in plant (photosynthesis, respiration, transpiration, crop duration, and phenology) are discussed in detail to realize the impact of climate change on yield and quality of cereals and other crops, field water balance, and water-use efficiency. It also elucidates how direct and interactive effects of climate variables are modified by photosynthetic pathway (C3 and C4) and N-fixing capability (legumes and nonlegumes) of the crop, water regime, and N level in the soil.

3.2 EFFECT OF CO_2 ENRICHED ATMOSPHERE ON SOIL ENVIRONMENT

Elevated atmospheric CO_2 alters soil environment by stimulating labile C pools, creating microbial diversity and changing nutrient status availability in soil. Rising atmospheric CO_2 concentration augments C pools above and below the ground by increasing plant biomass. The increased biomass, especially the shoot, enhances C allocation belowground as fine root growth and exudation (from decaying plants and roots). The increased root growth generates more C belowground, which helps in accelerating decomposition (depending upon C and N in litter) and root respiration (depending upon the availability of C and plant growth rates). Potential root growth and its surface area under higher CO_2 in soil are possible only if favorable soil physical conditions and composition of soil air, such as low mechanical impedance due to low bulk density and oxygen level of >2.1%, exist (Kirkham, 2011). In case the oxygen level falls down this level, stimulation of soil oxygen via tillage in arable lands and drainage in wet lands may become vital for root growth. The higher C stored in soil stimulates microbial activity and its biomass cycling, because the additional C (than the normal) is allocated exclusively to relatively labile pools such as microbial biomass carbon (MBC), readily mineralizable carbon (RMC), water-soluble carbohydrate carbon (WSC), acid-hydrolyzable carbohydrate carbon (AHC), and $KMnO_4$-oxidizable carbon ($KMnO_4$—C). The labile soil C pools turn over rapidly (≤ 2.0 years) with tillage, residue management practices, and climate, resulting in little net C storage (Taneva, Pippen, Schlesinger, & Gonzalez-Meler, 2006), and alter the availability of organic C, N, and cation nutrients for microbial populations in the soil. In many cases, increased C allocation to the root zone potentially alters the composition of root exudates ensuing change in the C:N ratio, availability of chemoattractants or signal compounds, microbial diversity, and nutrient availability. These changes may be negative, positive, or unchanged because of their extraordinary complexity (Diaz, Grime, Harris, & McPherson, 1993; Grayston, Campbell, Lutze, & Gifford, 1998; O'Neill, 1994; Prior et al., 1997; Randlett, Zak, Pregitzer, & Curtis, 1996; Schortemeyer, Hartwig, Hendrey, & Sadowsky, 1996; Williams, Rice, & Owensby, 2000).

Elevated CO_2 impacts beneficial bacterial communities and rhizospheric and endophytic populations, associated with plants. It prompts *rhizobia*, well known for their N fixation and for their additional plant-growth-promoting activities in legumes, and *Pseudomonas* spp., actinobacteria and deltaproteobacteria in nonlegumes. The dominance of microorganisms under variable environmental conditions is plant-specific. For example, the dominance of *Pseudomonas* sp., known to include many plant-growth-promoting members, is associated with rye, while *Rhizobium* sp. allied with white clover are boosted with increased CO_2 (Marilley, Hartwig, & Aragno, 1999). The plant specificity is of applications in agriculture or phytoremediation. For instance, with the increase of CO_2, the proportion of HCN-producing *Pseudomonas* strains, considered as potential inhibitors of root parasitic fungi, in two perennial grassland systems (*Lolium perenne* and *Medicago coerulea*) are reduced (Tarnawski, Hamelin, Jossi, Aragno, & Fromin, 2006). However, the proportion of siderophore producers and nitrate-dissimilating strains are increased. As increased CO_2 promotes beneficial bacteria, which are plant-specific, thus, it becomes important to identify plant-specific microbial strain that performs well under the altered climate conditions. For instance, plant-growth-promoting *Pseudomonas mendocina* strain enhanced the growth of lettuce plants under elevated CO_2 conditions (Kohler, Hernandez, Caravaca, & Roldan, 2009). Therefore, use of plant-growth-promoting bacteria (PGPB) is advocated to alleviate atypical stresses possibly imposed by climate change in the future under increased CO_2 concentrations.

Under elevated CO_2, changed soil C availability and microbial diversity alter the nutrient availability for growing plants. Plants take nutrients from the soil solution pool where nutrients are capable to move dominantly by mass flow, which is a function of physicochemical parameters of the soil, location of the ion relative to the root surface and length of the pathway of nutrient to be traveled in the soil to reach the root surface (Jungk, 2002). The bulk or mass flow of dissolved nutrient may be written formally as Eq. (3.1).

$$J_c = J_w C_i \qquad (3.1)$$

where J_c is nutrient flux (mass of nutrient per unit area per unit time), J_w is water flux (volume of water flowing per unit area per unit time), and C_i is total solute concentration (mass of solute per unit of soil volume). Higher concentration of atmospheric CO_2 increases the C content in soil that improves soil structure and helps soils to retain water and hold plant available nutrients, which serve as an energy source for decomposing organisms and decreases nutrient availability to plant. For example, in case of N, elevated CO_2 levels enhance plant-microbial N competition in the rhizosphere by increasing the root exudation and a related stimulation of rhizosphere-microbial growth and consequently lower N nutritional status of the plants. With an increase in C content, the C:N:P ratio is also changed, which may decrease the N and phosphorous (P) availability to plants.

Although elevated CO_2 decreases N content (percent) in plant tissues, N accumulation in canopy would increase slightly because of the enhancement in dry matter production with increased CO_2. Likewise, phosphorus (P) and potassium (K) uptake would also increase in aboveground tissues. Thus, increased CO_2 in atmosphere will increase in N-use efficiency for biomass production, grain output, and harvest index (HI) as the plants utilize the available N resources efficiently because of reduced critical N concentration for maximum production and less NO_3 leaching (Torbert, Prior, Rogers, & Runion, 2004). However, with high level of N fertilization under elevated CO_2, there is an increase in N content, but N-use efficiency would decrease. In the literature, controversial views exist about the role of enriched CO_2 on nutrient availability. Some expect elevated CO_2 to increase nutrient availability because of increased belowground C that will, in turn, enrich microbial C (Finzi, Sinsabaugh, Long, & Osgood, 2006), while others reported a decrease (Zhu & Cheng, 2011) and no change (Cheng & Kuzyakov, 2005). Reviewing literature, Pendall et al. (2004) suggested that increased CO_2 may not exert a significant direct effect on N mineralization per se, but associated warming can cause increased N mineralization, leading to increased solution-phase N. However, under limited microbial activity, N available to growing plants could decrease as N remains unavailable until broken down by soil microbial system in the soil. The availability of N for plant uptake under elevated CO_2 can be increased by reinforcement of microbial N mining mechanisms of increasing gross N mineralization rate in the rhizosphere (Koranda et al., 2011), which may eventually lead to higher soil N availability for root uptake due to faster turnover of microbes compared with roots (Kuzyakov & Xu, 2013) and increasing enzyme synthesis (Phillips, Finzi, & Bernhardt, 2011). Other nutrients, such as K, whose availability is not strongly controlled by biological activity, impacts of elevated CO_2 will be indirectly mediated by temperature and moisture changes.

3.3 EFFECT OF CO_2 ENRICHED ATMOSPHERE ON PLANTS

Carbon dioxide is the primary raw material utilized by plants to produce the organic matter, out of which they construct their tissues. Therefore, with a higher level of CO_2 in the air, plants can grow bigger and faster provided that levels of soil water, sunlight, temperature, and plant nutrients are adequate. Increased CO_2 may bring changes at different organization levels in plant biology, that is, gene expressions, cellular, organ, individual plant, and ecosystem. Detailed discussion on the effect of CO_2 at all the organization levels is beyond the scope of this book due to the complexity of the subject. However, responses of some of the important physiological processes like photosynthesis, respiration, and transpiration at canopy level to elevated CO_2 are discussed to comprehend the climate change impact on crops.

3.3.1 Photosynthesis

Photosynthesis is a process in which plants synthesize carbohydrates by using atmospheric CO_2 and plant water in the presence of light and chlorophyll following Eq. (3.2). This process also requires P and N:

$$6CO_2 + 6H_2O \xrightarrow[\text{Chlorophyll}]{\text{Light}} C_6H_{12}O_6 + 6O_2 \tag{3.2}$$

The synthesized carbohydrates are translocated to different parts of the plants. A part of the translocated carbohydrates gets utilized in different plant physiological processes, and the rest is stored in different plant parts and economic yield. Under elevated CO_2, stomatal apertures are partially closed; however, photosynthetic rate under such conditions is increased because, firstly, the internal CO_2 in the leaves remains more than sufficient and diffusion of CO_2 to the fixation site continues for carboxylase fixation; secondly, leaf duration is lengthened; thirdly, optimum temperature for photosynthesis is shifted upward; and fourthly, water stress is reduced by stomatal closure. It has been found that photosynthetic rate increases about three times with a four times increase in CO_2 concentration (Gaastra, 1962). Although photosynthesis by elevated CO_2 is stimulated in the short term, overtime photosynthetic rate declines inevitably. This process is termed photosynthesis acclimation. The photosynthesis acclimation or decrease in photosynthetic ability of the plant occurs due to four reasons, (i) the loss of photosynthetic components; (ii) decreased activation of components; (iii) reduced water flow through the plant due to its physiological effect of closing stomata; and (iv) less N uptake due to increased C:N ratio. Acclimation mostly occurs in plants grown under unnatural environment, that is, low N level and water stressed conditions, where root growth is restricted. In tubers, the beneficial effect of CO_2 is less due to more acclimation during biomass production and tuber formation; consequently, the active stimulation of photosynthesis is decreased during the period of active tuber filling. When water and nutrient supplies are sufficient and temperatures are warm, the response of increased CO_2 to photosynthesis is positive, even in young plants. Actually, such conditions stimulate extra root growth and N uptake. The concentration of CO_2 beyond which acclimation phenomenon starts is crop-specific. Photosynthesis acclimation may occur with long-term exposure to elevated CO_2 beyond 1000 ppm in rice and 800 ppm in soybean.

Physiological response of CO_2 also varies with species and photosynthetic pathway (Torbert et al., 2004). Based on photosynthetic pathway, plants are divided in three groups, namely, C3, C4, and Crassulacean acid metabolism (CAM). C3 and C4 plants are designated based on the number of C atoms involved in the compound of initial products of the photosynthesis. In C3 plants, the initial product of photosynthesis is phosphoglyceric acid (containing three C atoms) but in C4 is malic acid or aspartic acid (containing four C atoms). The crops using C3 pathways are cereals (wheat, rice, barley, oats, rye, triticale, etc.), legumes (dry bean, soybean, peanut, mung bean, faba bean, cowpea, common pea, chickpea, pigeon pea, lentil, etc.), fruits (including banana and coconut), roots and tubers (potato, taro, yams, sweet potato, cassava, and sugar beet), fiber crops (cotton, jute, sisal, etc.), oil crops (sesame, sunflower, rapeseed, safflower, etc.), and trees. The crops using C4 pathway are maize, sorghum, millets, and sugarcane. CAM crops are cacti, succulents, agaves, and pineapple. The C3 and C4 photosynthetic pathways are also differentiated due to the level of CO_2 saturation of ribulose bisphosphate carboxylase-oxygenase (rubisco) in plant cells. Rubisco is an enzyme in chloroplasts that fixes atmospheric CO_2 during photosynthesis. The level of CO_2 saturation in rubisco determines CO_2 utilization in photosynthesis. In case rubisco is CO_2 saturated, the total CO_2 is utilized in photosynthesis. If not, CO_2 undergoes another reaction with oxygen mediated by similar rubisco enzyme and produces phosphoglycoate, which metabolizes and releases CO_2 in photorespiration process. At current level of atmospheric CO_2 conditions, rubisco in C3 plants is not CO_2-saturated; therefore, CO_2 undergoes in photorespiration process and releases CO_2. Because of the released CO_2, the net rate of photosynthesis is decreased. In other words, C3 photosynthesis is less efficient than C4 partly because of photorespiration, which results in the loss (to the atmosphere or soil) of a substantial proportion of the C that has been extracted from the atmosphere by photosynthesis. In C3 plants, photorespiration is more under heat stress and drought conditions and can be suppressed with increased CO_2, which enhances photosynthesis by promoting carboxylation and diminishing oxygenation of photosynthetic enzyme, Rubisco. On the other hand, in C4 plants, Rubisco remains saturated by CO_2 even at comparatively low CO_2 levels in the leaf tissue due to the presence of phosphoenolpyruvate, the primary acceptor of CO_2 having more affinity with CO_2. For example, maize being a C4 plant can accommodate 10–100 times more CO_2 than that of C3 plants (Furbank & Hatch, 1987). Due to more of CO_2 (three to six times more than the atmospheric CO_2) in C4 plants, an increase in external CO_2 has little or no effect on net photosynthesis as Rubisco is already saturated at current level of CO_2 (Von Caemmerer & Furbank, 2003). The C4 plants respond to elevated CO_2 only in the conditions that CO_2 is leaked from bundle sheet cells due to N deficiency and drought stress (Furbank & Hatch, 1987). That is the reason C4 crops are preferred for hot dry climates at ambient level of CO_2. Sometimes younger leaves show sensitivity to CO_2 concentration due to low N uptake (Cousins et al., 2003). The response of increased CO_2 to photosynthesis in young plants becomes positive when temperature is warm and N supply is not limiting. Probably such conditions stimulate root growth and N uptake. Photosynthesis is also stimulated in CAM species by high CO_2. The stimulation depends upon the ability to switch over to C3 pathway when water is available, that is, early and late in the day. So, depending upon photosynthetic pathway, C3, C4, and CAM, photosynthesis in plants respond differently to elevated level of CO_2. For example, the benefit of doubling CO_2 (550 ppm) to increase photosynthesis was 30%–50% in C3 and 10%–25% in C4 plant species (Ainsworth & Long, 2005; Gifford, 2004; Long, Ainsworth, Rogers, & Ort, 2004).

Briefly, the increased CO_2 level at the site of Rubisco either metabolically (as in C4) or abiotically (as increased atmospheric CO_2) reduces rate of oxygenation and suppresses photorespiration with a subsequent increase in net photosynthetic rate. At present levels of CO_2, C4 plants are more efficient (7%) at photosynthesis than C3 (4%) (https://buythetruth.Wordpress.com). In the future, with rise in CO_2 level, photosynthetic efficiency gap between C3 and C4 plants would narrow down, and the photosynthesis rate in C3 plants will be as good as or better than C4 plants, and the optimum temperature for photosynthesis would be higher (Fig. 3.1).

In photosynthesis, process plants also require nutrients along with CO_2 and water to synthesize carbohydrates. There is an interaction of CO_2 and N, which affects the photosynthesis. Less enhancement of photosynthesis by elevated CO_2 has been observed under low than high soil N conditions in FACE experiments (Ainsworth & Long, 2005; Ainsworth & Rogers, 2007). The interaction of N supply and CO_2 is also influenced by photosynthetic pathway of plants.

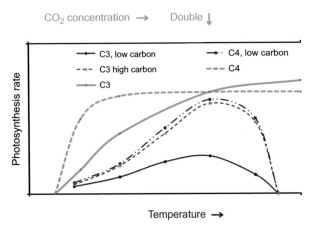

Fig. 3.1 Schematic diagram of photosynthesis rate with increased CO_2 concentration and temperature in C3 and C4 plants.

In C3 plants, increased CO_2 increases N uptake resulting from better photosynthesis and biomass, while no or less increase in N uptake in C4 plants. Therefore, nutrient requirements in C4 plants are not expected to increase under elevated CO_2. In fact, the photosynthetic N–use efficiency in C4 plants at present level of CO_2 concentration is more (almost double) than that of C3 plants (Long, 1999). Among C3 species, legumes respond more as compared with nonlegumes to elevated CO_2 and ultimately increase photosynthesis (Rogers, Ainsworth, & Leakey Andrew, 2009). Actually, legumes maximize the benefits of elevated atmospheric C by pushing excess C to root nodules where it (i) serves as a C and energy source for the bacterial symbionts and (ii) exchanges the excess C for N. Thus, under elevated CO_2, not only photosynthesis is higher in legumes, but also the decrease in tissue N concentration is also smaller than for other C3 species (Cotrufo, Ineson, & Scott, 1998; Jablonski, Wang, & Curtis, 2002; Taub, Miller, & Allen, 2008). In FACE experiments, soybean showed a greater response to elevated CO_2 than wheat and rice in photosynthesis and overall growth, possibly because of less N decrease in plant tissues in the former. However, response to harvestable yield was insignificant (Long, Ainsworth, Leakey, Nosberger, & Ort, 2006). Under elevated CO_2 environment, photosynthesis is also altered due to the decreasing effect on respiratory component. Increased CO_2 suppresses photorespiration (due to the oxygenation of photosynthetic enzyme, Rubisco), increases dark respiration (due to carboxylase fixation), decreases specific respiration per unit mass (due to the reduced N concentration of aboveground biomass), and lowers night respiration. These changes in respiration along with increased CO_2 diffusion gradient into leaves lead to enhanced photosynthetic rates.

3.3.2 Stomatal Conductance and Water-Use Efficiency

On the underside of plant leaves, there are tiny holes, called *stomata*, which are the openings through which plant absorbs CO_2. With higher level of CO_2 concentration in the air, the stomata do not have to be open as wide and remain narrower to receive the required amount of CO_2. As a result, the stomatal resistance is increased, or stomatal conductance to water vapors is decreased. It (decreased stomatal conductance) decreases transpiration and increases leaf water-use efficiency (WUE), usually, *defined as ratio of leaf carbon uptake to water loss.* Narrowing the stomatal aperture of plants with increased CO_2 has the additional benefit that it restricts the entry of pollutants present in the air such as ozone or sulfur dioxide. Leaf level transpiration rate (T_{rL}) can be expressed as Eq. (3.3):

$$T_{rL} = K\left[e_{T_L} - e_a \big/ r_a + r_s\right] \qquad (3.3)$$

where T_{rL} is in $mmol\,m^{-2}\,s^{-1}$, K is a temperature-dependent physical constant for conversion of vapor pressure (kPa) to gas concentration ($mmol\,mol^{-1}$), e_{TL} is saturation vapor pressure at leaf temperature (kPa), e_a is vapor pressure of air (kPa), r_a is resistance of boundary layer, and r_s is stomatal diffusive resistance of water vapors ($mol\,m^{-2}\,s^{-1}$). In Eq. (3.3), increased CO_2 affects the stomatal resistance and vapor pressure gradient. Increased CO_2 though increases stomatal resistance and decreases transpiration, but a decrease in leaf transpiration is relatively less, because the effect of decreased stomatal conductance due to stomatal closure was offset by the effect of a concomitant increase in leaf temperature resulting from a decrease in transpiration cooling (Burkart, Manderscheid, Wittich, Lopmeier, & Weigel, 2011). The increase in foliage temperature could be 1–2°C (Prasad, Boote, Allen, & Thomas, 2002; Pan, 1996). The increased leaf temperature increases saturated vapor pressure in leaf and vapor pressure gradient between leaf and air. It means that the effect of elevated CO_2 on transpiration depends not only on stomatal resistance in leaves but also on evaporative demand of the environment, which in turn is determined by the factors such as temperature, solar radiation, RH, and wind speed. This process has been described by Penman-Monteith equation (Monteith, 1965), which defines that transpiration is affected through growth and leaf area, change in stomatal aperture and conductance for water vapor loss, and vapor pressure gradient between the ambient air and substomatal cavity. It is used to calculate potential evapotranspiration, ET_0 (as Eq. 3.4):

$$ET_0 = \frac{0.408\Delta(R_n - G) + \gamma\frac{900}{T + 273}u_2(e_s - e_a)}{\Delta + \gamma(1 + 0.34u_2)} \tag{3.4}$$

where Δ is the slope of the saturation vapor pressure at mean air temperature curve (kPa $°C^{-1}$), R_n and G are the net radiation and soil heat flux density in $MJ\,m^{-2}\,day^{-1}$, γ is the psychrometric constant (kPa $°C^{-1}$), T is the daily mean temperature (°C), u_2 is the mean wind speed in $m\,s^{-1}$, e_s is the saturation vapor pressure (kPa) calculated from the mean air temperature (°C) for the day, and e_a is the actual vapor pressure (kPa). Potential crop ET is determined by multiplying ET_0 by a crop coefficient (Kc).

The decrease in stomatal conductance (or increase in stomatal resistance) under elevated than ambient CO_2 conditions is inconsistent and varies with experimental conditions and photosynthetic pathway in plant and water and N supply from soil. In different experimental conditions, the decrease in stomatal conductance was to the tune of 77%–86% in many species (Bunce, 1995), $40 \pm 5\%$ from analysis of 86 sets of experiments (Morison, 1987), 30%–40% in wheat crop (Samarakoon, Muller, & Gifford, 1995), 40% in soybean (Ainsworth et al., 2002; Ainsworth & Rogers, 2007), 22% across a number of FACE experiments (Ainsworth & Rogers, 2007), and 34% from chamber experiments (Kimball & Idso, 1983). The reduction in stomatal conductance with elevated CO_2 is more in C3 than C4 species. A meta-analysis on wild grasses reveals that elevated CO_2 reduced stomatal conductance by 39% in C3 and 29% in C4 species (Wand, Midgley, Jones, & Cuttis, 1999).

The decrease in stomatal conductance is because of partial closure of stomatal aperture, not due to stomatal density (stomata/unit leaf area) and stomatal index (stomata/leaf epidermal cell) (Morison, 1998). The decrease in water conductance by stomata is more in water stressed or drought conditions than in watered treatment and with zero N than with adequate N (Li, Kang, & Zhang, 2004).

Decreased stomatal conductance by high CO_2 decreases ET, but the decrease in ET is less than stomatal conductance (Field, Jackson, & Mooney, 1995). The decrease in canopy ET varied from different crops in different sets of experiments depending upon the decrease in stomatal conductance and increase in leaf area under elevated CO_2 and prevailing evaporative demand. For instance, the decrease in ET was 9%–12% in soybean (Bernacchi, Kimball, Quarles, Long, & Ort, 2007), 10% in rice (Yoshimoto, Oue, & Kobayashi, 2005), and 8% in wheat (Andre & Du Cloux, 1993; Kimball et al., 1999). In addition to crops, soil water regimes and N levels also affect ET under the elevated CO_2 as there is a significant interaction between $CO_2 \times$ soil water $\times$ N (Li et al., 2004). The interaction indicates that by doubling CO_2 (from 350 to 700 ppm), the decrease in canopy ET started at higher level of N ($337.5 \, kg \, ha^{-1}$) under well-watered conditions and at lower N level ($112.5 \, kg \, ha^{-1}$) under drought conditions. Under adequate water and N supply, the increase in ET brought out by higher leaf area overweighs the CO_2 enrichment-induced decrease in transpiration. Such interactions are of great importance for areas where water and N availability are main constraints. Under limited water and N supply, CO_2 enrichment has limited effect on either leaf area or ET. Under increased N and water supplies, leaf area and ET are increased, but there is an inability of plants to increase leaf area due to increased CO_2. From studies in Ohio and Florida, Boote, Pickering, and Allen (1997) reported that with increasing CO_2 from 350 to 700 ppm, transpiration was decreased 11%–16% for irrigated sites and 7% for rain-fed, while ET was decreased from 6% to 8% for irrigated sites and 4% for rain-fed site. In crops where elevated CO_2 stimulates an increase in leaf size relatively more than stomatal closure, no reduction in transpiration can be expected in canopies (Samarakoon & Gifford, 1995). Actually, CO_2 affects transpiration in plants by two ways. First, CO_2 directly controls the stomatal conductance through its control on stomatal aperture. Second, CO_2 promotes growth by increased photosynthesis, thus producing more leaf area that will increase transpiring surface area per unit of land area. The second way of affecting water-use rates would be important before the crop achieves complete ground cover. After this, the effect of greater leaf area index (LAI) can override the effect of decreased stomatal conductance resulting from elevated CO_2. Due to these two effects (low stomatal conductance and increased leaf area), three types of responses of ET to elevated CO_2 are expected. Increased CO_2, firstly, decreases ET due to a decrease in stomatal conductance; secondly, increases ET with increased leaf area in crop plants (Kellither, Leuning, Raupach, & Schelze, 1995); and thirdly, results in no or very little change in ET due to little increase in leaf area under limited water and N supply or as a result of photosynthesis acclimation in crops (Bunce, 2000). ET response to elevated CO_2 is also modified by atmospheric temperature. At high temperature, response of elevated CO_2 to reduce ET is less. For example, with doubling CO_2 in soybean at temperature 20/18°C, ET was reduced by 9% and no reduction at 40/30°C (Allen, Pan, Boote, Pickering, & Jones, 2003). In fact, ET increases at higher temperature due to diminishing of CO_2-induced benefit of decreased stomatal conductance with an increase in temperature (Horrie, Baker, Nakagawa, Matsui, & Kim, 2000).

Although with increased CO_2, WUE (defined as *amount of dry matter/ yield produced per unit of water lost in evapotranspiration from the same area of ground*) is increased in all plants, yet the increase is more in C3 than C4 plants at elevated CO_2. By summing up 46 observations, Kimball and Idso (1983) revealed a doubling of WUE in C4 crops (maize and sorghum) for a doubling of CO_2 level. In controlled experiments in glass house, Samarakoon et al. (1995) found 64% increase in WUE in C3 crop (wheat) by doubling CO_2. The increased WUE with increased CO_2 may be either an increase in yield by photosynthesis or a decrease in water loss by evapotranspiration when compared with ambient CO_2 (Conley et al., 2001); however, their contributions may vary with photosynthetic pathway. The increased photosynthesis contributes more in C3 plants than C4, while reduced transpiration contributes more in C4 than C3 (Samarakoon & Gifford, 1996). Acock and Allen (1985) using data from Valle, Mishoe, Jones, and Allen (1985) and Wong (1980) have also demonstrated that WUE in C4 plants is primarily controlled by transpiration only, whereas both, photosynthesis and transpiration, are important in C3 plants. Quantification of the contributions of photosynthesis and transpiration to WUE in C4 and C3 plants by Prior et al. (2010) demonstrates that photosynthesis contributes 74% in soybean (C3) compared with 42% in sorghum (C4). In the corresponding crops, the contributions of transpiration were 26% and 58%. In some cases, less transpiration decreases mass flow of nutrients to roots, changes soil moisture pattern, and increases foliar temperature, which reduces photosynthesis and WUE. Moreover, enhancement of photosynthesis under elevated CO_2 in C3 plants is direct but in C4 indirect, that is, by avoiding salinity and water stresses. That is why the increase in WUE with increased CO_2 is more in C3 plants than C4 (Kimball, 1983). With increased CO_2 (ambient + 200 ppm), there was an increase in WUE by 33%–40% in C3 and 10%–15% in C4 plants (Prior, Torbert, Runion, & Rogers, 2003). The effect of increased CO_2 on WUE is also influenced by the soil water content, leaf area, and N supply from soil. Under adequate soil moisture conditions, elevated CO_2 had no effect on WUE; however, the effect is more under water stressed conditions, and that is through more effective and prolonged use of soil water by reducing water loss as transpiration plus maintaining higher water in the plant body. In effect, reduced water loss by transpiration under elevated CO_2 increases whole-plant hydraulic conductance and osmotic adjustment, which lead to conditions whereby plants maintain higher (less negative) leaf water potentials, creating the same effect as providing more water. That is why it has been often pointed out that effect of drought can be negated a bit by elevated CO_2. Under elevated CO_2 of 550 ppm, WUE in drought conditions compared with well-water conditions was more by 22% (Manderschei & Weigel, 2000). Similarly, in a simulation study with increasing CO_2 from 350 to 700 ppm in soybean, WUE was increased by 53%–61% (Boote et al., 1997), closer to 50%–60% reported by Allen et al. (2003). Elevated CO_2 increases WUE for plants with both high and low LAI, but this increase was greater for plants with a lower LAI (Jones, Allen, Jones, & Valle, 1985) because of relatively lower magnitude of ET with stomatal closure. Enhanced WUE gets offset when ET increased with increased leaf area is counterbalanced with the ET reduced due to decreased stomatal conductance (Allen, 1994). Nitrogen level in soil and atmospheric temperature also influence the increase in WUE at elevated CO_2. The increase in WUE is comparatively higher at high N due to more photosynthesis than under low N level. The increased WUE with increased CO_2 declines sharply as the temperature exceeds optimum. At temperature higher than the optimum, WUE is decreased due to more reduction in yield than that of ET (Buttar, Jalota, Sood, & Bhushan, 2012). When both CO_2 and temperature are higher, the resultant WUE is determined by the overweighing effect of the temperature, which causes more reduction in yield than ET.

3.3.3 Crop Yield

Elevated levels of anthropogenic CO_2 are beneficial to plants in a process, which is described as CO_2 fertilization and is commonly defined as *enhancement of the net primary production by CO_2 enrichment as a result of an increase in atmospheric concentration.* The additional photosynthate availability in elevated CO_2 enables most plants to grow faster and yield more. The more crop yield with increased CO_2 is accredited to the favorable changes from seed germination to reproduction stages in the plants. Increased CO_2 enhances seed germination, nodes and branches, tiller formation, leaf formation and root growth (root length, root diameter and cortex width, fine root colonization of arbuscular mycorrhizal fungi, and nodule formation), floral number and pollen production, grain/seed/fruit size, and percentage of filled grains.

Increase in potential yield due to these favorable changes with elevated CO_2 is possible only under potential irrigated and fertilized conditions. Elevated CO_2 exerts its positive effects, firstly by enhancing crop growth and WUE; secondly by enabling the plants to better withstand the growth retarding effects of environmental stresses, including high and low temperatures, salinity, air pollution, and airborne and soilborne pathogens; and thirdly by improving yield attributes. The first effect is due to increased photosynthesis and decreased stomatal conductance. The second effect is via modification of water and nutrient cycles, improving soil-water-plant relations through changes in physical properties of soil (soil C, aggregate stability, hydraulic conductivity, water infiltration, bulk density, etc.) and increased plant rooting. The third effect is via increased biomass, root-shoot ratio, leaf area, panicles/plant, panicle dry weight, percentage of filled grains, the number of tillers, number of grains/spike, HI, and panicle growth rate in cereal crops like rice and wheat; root and shoot length, leaf areas, root volume, the number of pods, 1000 grain weight, higher number of pods per plant, more seeds/pods, and HI in pod crops like black gram; and the number of bolls/plant, boll dry weight, and lint dry weight/boll in fiber crops like cotton. Increased CO_2 also increases the N uptake, which increases fertile spikelets, individual grain weight, and percentage of filled grains.

A combination of these responses of higher CO_2 level would mean that all major crops can grow better and yield more (Hatfield et al., 2011; Izaurralde et al., 2011). Biomass and yield tend to increase significantly in all plants, especially of C3 plants, as CO_2 concentrations increase above current levels. Dry matter produced under elevated CO_2 is though more for belowground portions of plants than for the aboveground portion (Ainsworth & Long, 2005; De Graaff, Van Groenigen, Six, Hungate, & Van Kessel, 2006), yet the increased plant growth reflected in the harvestable yield of wheat, rice, and soybean, all showing yield increase of 12%–14% (Ainsworth, 2008; Long et al., 2006). In northwestern part of India, rice and wheat showed an increase in yield by 15% and 28%, respectively, at elevated CO_2 concentrations of 660 ppm (Lal, Singh, Rathore, Srinivasan, & Saseendran, 1998). Rice yield under elevated CO_2 (570 ppm) was 24% higher than under ambient CO_2 concentration (370 ppm) in subtropical regions (De Costa, Weerakoon, Herath, Amaratunga, & Abeywardena, 2006). No doubt, there are benefits of higher CO_2 on photosynthesis, water-use efficiency, and symbiotic fixation of N in leguminous crops and yield in all crops, but the mode by which elevated CO_2 increases yield is crop and photosynthetic pathway-specific. In C3 crops, elevated CO_2

increases yield by contributing dry matter to pods during reproductive stage in soybean, by increasing biomass in rice, and by faster growth and more interception of radiation in cotton. In C4 plants, elevated CO_2 is generally assumed to have little or no effect on yield, but this assumption does not hold true, particularly for rain-fed crops (Samarakoon & Gifford, 1996). This has been supported by a significant increase in biomass and yield of sorghum for water deficit treatments and not significant when grown with full irrigation at the Arizona FACE project (Ottman et al., 2001). But such effects are realized only when water-sensitive growth stages, especially flowering and grain filling are not disrupted by some environmental stress like drought or high temperature.

The magnitude of benefit of elevated CO_2 could be not so far as when combined with other climatic parameters like temperature and P_{cp} and fertilizer N application, the main yield determining edaphic factors. The effect is also modified by insect attack and weed competition. At elevated CO_2, irrigation and fertilizer applications add to the beneficial effect of CO_2, especially under low-temperature conditions (Li, Han, Zhang, & Li, 2007). High temperature negates the positive effect of elevated CO_2 by shortening crop duration, especially grain-fill period, low/high P_{cp} by posing water/aeration stress, and low N by checking crop growth. Smaller response of rice yield to elevated CO_2 under low (growth limiting) N rates has been observed by Kim et al. (2003). The beneficial response of elevated CO_2 to yield on account of lengthening of crop duration may get abridged (i) due to an increasing number of insect generations per year or invited migratory insects; (ii) by increasing competition for the same resources (light, water, CO_2, nutrients, etc.) on the same patch of land due to stronger growth response of many C3 weeds to increasing CO_2 than most cash crops (Ziska, Reves, & Blank, 2005); and (iii) by decreasing efficacy of the herbicides because of less absorption and translocation with partially stomatal closure (Ziska & Teasdale, 2000). This means that efficiency of elevated CO_2 could diminish in agricultural systems where efficient measures to control weeds and insects are not practiced. In mixed-species experiments of C3 and C4 plants, the benefit of elevated CO_2 depends upon the interaction of photosynthetic pathway and N-fixing capability of the plant and the soil N fertility. Among C3, the benefit is more in nonlegumes under high-fertility conditions and in legumes under low-fertility conditions (Porter & Navas, 2003). In general, the benefit is more in C3 plants than C4 under high-fertility conditions. The effect of CO_2 modified by temperature and P_{cp} and use of N fertilizer have been found in results across a variety of experimental settings, that is, controlled environment closed chambers, greenhouses, open and closed field top chambers, and FACE experiments. But most of the plant physiologists and modelers are of the view that the effects of CO_2, measured in controlled experimental setting or subsequently implemented in models (Tubiello et al., 2007), may overestimate actual field- and farm-level responses because effects of many limiting factors such as pests and weeds, nutrients competition for resources, soil, water, and air quality have not been adequately taken into account in the major models.

Over and above these affirmative and pessimistic effects of elevated CO_2 on crop yields, it (increased CO_2) deteriorates the quality of marketable yield of the crops. Due to the robust CO_2 and N interaction (Kimball et al., 2001; Sinclair et al., 2000), N content in plant tissues is decreased under elevated CO_2 environment, which may occur through (i) dilution of N from increased carbohydrate concentrations and (ii) less water uptake owing to a decrease in stomatal conductance. The decreased N diminishes the rates of assimilation of nitrate into organic compounds (Bloom, Burger, Rubio Asensio, & Cousins, 2010) and nitrate photoreduction (Rachmilevitch, Cousins, & Bloom, 2004). Decreased N in plants either from increased atmospheric CO_2 or with applied N fertilizers at low rates reduces protein content in grains. With a decrease in N to half of the recommended dose, there was a reduction of crude protein by 4%–13% in wheat and 11%–13% in barley and an increase in starch in both species by 4% (Erbs et al., 2010). In FACE experiments, due to low N content under elevated CO_2 of 540–958 ppm compared with ambient (315–400 ppm), protein concentrations were less by 10%–15% in wheat, barley, and rice (Taub et al., 2008). For soybean, the reduction in protein concentration was 1.4%, and for potato, it was 14%. Evidence of declining of protein content with rising CO_2 in forage species is also documented (IPCC, 2001). The decrease in crude proteins, total and soluble β-amylase, in the grains diminishes the nutritional and processing quality of wheat flour by affecting kernel hardness. The concentrations of nutritionally important minerals including calcium, magnesium, and phosphorus in crops also decreased under elevated CO_2 (Loladze, 2002; Taub & Wang, 2008). At elevated CO_2 and temperature, grain starch content in winter wheat is less (Tester et al., 1995).

At global level, CO_2 is a critical factor for producing crop yields (Nelson et al., 2009; Parry, Rosenzweig, Iglesias, Livermore, & Fischer, 2004). Increased productivity by increasing CO_2 fertilization would be shown in the driest regions. North America and Europe may be benefitted from the increased CO_2 fertilization for a short run, but other regions such as Africa and India may not be benefitted even with strong fertilization due to warming effect. Results showing evidences of increased yield of different crops with an increase in CO_2 (from 330 to 760 ppm) under variable experimental conditions are presented in Table 3.1. The plant response to changes in CO_2 concentration is intricate as it is modified by temperature, soil moisture, and nutrient management and their interactions. A linear relation fitted to the data of Table 3.1 indicates that the percent increase in yield with the percent increase in CO_2 is more in C3 plants than C4 (Fig. 3.2). At 100% increase in CO_2, the increase in yield was 39% in C3 and 24% in C4 crops.

Other researchers (Ainsworth & Long, 2005; Gifford, 2004; Long et al., 2004) have also reported an increase in crop yield in the range of 10%–20% for C3 crops and 0%–10% for C4 crops at 550 ppm CO_2 compared with current atmospheric CO_2 concentrations of 380 ppm. However, in general doubling, CO_2 (from 330 ppm) increased yield by 30% and 10% in C3 and C4 crops, respectively (Hatfield et al., 2011). In fact, under increased CO_2 concentration, C4 plants have less efficiency to increase photosynthesis that determines the yield. So, the benefits of increasing photosynthesis, symbiotic fixation of N in legumes, resistance to lower temperatures and air pollution, and tolerance of soil and water salinity by the higher CO_2 help increase WUE and yield of crops in the areas considered today too arid or where precipitation (P_{cp}) is going to decline.

3.4 EFFECT OF RISING AIR TEMPERATURE ON SOIL ENVIRONMENT

Rising temperature increases the rate of chemical and biochemical reactions in soil and plant. The increase in rate of reaction by the increased temperature is expressed most simply in terms of a temperature coefficient, Q_{10} value, derived from Eq. (3.5):

$$Q_{10} = \left(\frac{K_1}{K_2}\right)^{10/(t_1 - t_2)}$$

(3.5)

where K_1 and K_2 are velocity constants (proportional to the rates of reaction) found at temperatures t_1 and t_2. In simple terms, the Q_{10} for a particular reaction is an expression of the predicted increase in rate for a 10°C increase in temperature. For most biological reactions, the value of Q_{10} is between 2 and 3. In plants, growth rate becomes more than double by 10°C rise in

Table 3.1 Percent Increase in Yield of Different Crops With Change in CO_2 Concentration in Air

Author	Crop	Experimental Conditions/Model Used	Level of CO_2 (ppm)	Increase in Yield (%)
Experimental				
Schutz and Fangmeier (2001)	Spring wheat	OTC	650 (base 350)	46
Weigel et al. (1994)	Wheat (cultivar, star)	OTC	718 (base 384)	19
Weigel et al. (1994)	Wheat (cultivar, turbo)	OTC	718 (base 384)	27
Mulhollet al. (1997)	Spring wheat	OTC	680 (base 350)	33
Kimball et al. (1995)	Wheat (wet condition)	FACE	550 (base 370)	8
Kimball et al. (1995)	Wheat (dry condition)	FACE	550 (base 370)	20
Pandey et al. (2007)	Wheat		440, 550, 660 (base 330)	21, 48, 68
Kang et al. (2002)	Wheat	OTC	700 (base 350)	47
Tubiello and Ewert (2002)	Wheat	Field	700 (base 350)	32
Zhou et al. (2002)	Wheat	Field	400, 550, 600 (base 360)	7, 20, 30
Li et al. (2005)	Wheat	Field	450 (base 360)	21
Xiao et al. (2005)	Wheat	Field	450 (base 360)	16
Bender et al. (1999)	Wheat	OTC	Per 100 $\mu mol\,mol^{-1}$	11
Batts et al. (1997)	Wheat	OTC	Per 100 $\mu mol\,mol^{-1}$	17
Pinter et al. (1996, 1997)	Wheat	OTC	Per 100 $\mu mol\,mol^{-1}$	7
Weigel et al. (1994),	Barley (cultivar, Alexis)	OTC	718 (base 384)	52
Weigel et al. (1994)	Barley (cultivar, Arena)	OTC	718 (base 384)	89
Baker et al. (1992)	Rice	OTC	660 (base 330)	37
Baker et al. (1990)	Rice	OTC	660, 900 (base 330)	47, 74
De Costa et al. (2006)	Rice	OTC	570 (base 370)	24
Kim et al. (1996)	Rice	Field	–	30
Kang et al. (2002)	Cotton	OTC	700 (base 350)	38
Mauney et al. (1994)	Cotton	FACE	550 (base 280)	40

Reference	Crop	Model/Method	CO_2 (base)	Value
Reddy et al. (2004)	Cotton	OTC	520 (base 360)	4
Kimball and Mauney (1993)	Cotton	–	650 (base 350)	60
Vanaja et al. (2007)	Chickpea (biomass)	OTC	550, 700 (base 365)	39, 65
Torbert et al. (2004)	Sorghum	OTC	715 (360 base)	30
Prior et al. (2005)	Sorghum	OTC	683 (base 375)	40
Kang et al. (2002)	Maize	OTC	700 (base 350)	15
Prior et al. (2005)	Soybean	OTC	683 (base 375)	6
Baker et al. (1989)	Soybean	Field	660 (base 330)	33.6
Torbert et al. (2004)	Soybean	OTC	715 (base 360)	40
Simulations				
Lal et al. (1998)	Wheat	CERES-wheat	660 (base 330)	28
Jalota et al. (2009)	Wheat	CropSyst	760 (base 380)	7
Kaur and Hundal (2010)	Wheat	CERES-wheat	600 (base 330)	5.6
Cuculeanu et al. (1999)	Wheat	CERES-wheat	660 (base 330)	5–34
Brown and Rosenberg (1999)	Winter wheat	EPIC	560 (base 365)	18–29
Iglesias and Minguez (1997)	Wheat	CERES-wheat	617 (base 353)	5–85
Ghaffari et al. (2002)	Wheat	CERES-wheat	400–430 (base 370)	2–5
Southworth et al. (2002b)	Winter wheat	CERES-wheat	550 (base 360)	20–30
Hundal and Kaur (1996)	Rice	CERES-rice	400, 500, 600 (base 330)	1.5, 6.6, 8.7
Hundal and Kaur (2007)	Rice	CERES-rice	600 (base 330)	0.5
Jalota et al. (2009)	Rice	CropSyst	760 (base 380)	4.9
Lal et al. (1998)	Rice	CERES-rice	660 (base 330)	15
Saseendran et al. (2000)	Rice	CERES-rice	460 (base 360)	12
Hundal and Kaur (1996)	Rice	CERES-rice	400, 500, 600 (base 330)	1.5, 6.6, 8.7
Jalota et al. (2009)	Rice	CropSyst	760 (base 380)	4.9

Continued

Table 3.1 Percent Increase in Yield of Different Crops With Change in CO_2 Concentration in Air—cont'd

Author	Crop	Experimental Conditions/Model Used	Level of CO_2 (ppm)	Increase in Yield (%)
Reddy et al. (2002)	Cotton	GOSSYM	540 (base 360)	10
Jalota et al. (2009)	Cotton	CropSyst	760 (base 380)	5.5
Jalota et al. (2009)	Maize	CropSyst	760 (base 380)	6.5
Sahoo (1999)	Maize	CERES-maize	700 (base 350)	9
Southworth et al. (2000)	Maize (short duration)	CERES-rice	550 (base 360)	5–35
Cuculeanu et al. (1999)	Maize	CERES-maize	660 (base 330)	29–74
Brown and Rosenberg (1999)	Maize	EPIC	560 (base 365)	2–5
Iglesias and Minguez (1997)	Maize	CERES-maize	617 (base 353)	–2–16
Kaur and Hundal (2010)	Soybean	DSSAT	600 (base 330)	35.2
Lal et al. (1999)	Soybean	CROPGRO-soybean	660 (base 330)	50
Southworth et al. (2002a)	Soybean	SOYGRO	550 (base 360)	20
Kaur and Hundal (2010)	Groundnut	CERES-chickpea	600 (base 330)	30.4
Kaur and Hundal (2010)	Chickpea	DSSAT	600 (base 330)	55.7
Jalota et al. (2009)	Rice–wheat	CropSyst	760 (base 380)	7
	Maize–wheat	CropSyst	760 (base 380)	7
	Cotton–wheat	CropSyst	760 (base 380)	4

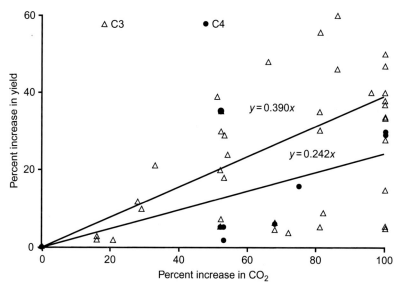

Fig. 3.2 Response of yields of C3 and C4 plants to increased CO_2.

temperature; however, the response is reduced at higher temperature and works within 20–30°C, which is optimum for many crops.

In the soil, an increase in temperature increases soil microbes, microbial respiration, and root respiration, which release CO_2. Higher temperatures also speed up the natural decomposition of organic matter, which reduces the C available/stored in soil, necessary for retaining water and important nutrients for biomass production in growing plants and serving as energy source for decomposing organisms. Increased temperature can increase N availability to plants through higher turnover of soil N. Although higher temperature increases N mineralization rate in soil, its recovery may diminish due to increased gaseous losses through processes such as volatilization and denitrification. Temperature and soil moisture being primary determinants of nutrient availability influence nutrient-use efficiency through direct impacts on root surface area. In effect, temperature in combination with soil moisture increases root growth and C allocation to roots that govern nutrient acquisition. Increased temperature increases the process of N fixation, associated with greater root development with higher CO_2, if soil moisture is not limiting. The continual cycling of plant nutrients, that is, C, N, P, K, sulfur, etc. in the soil-plant-atmosphere system is also likely to accelerate in warmer conditions. Elevated temperature influences soil water by increasing evaporation rate from the soil and accelerating transpiration rate in the plants, which causes moisture stress. Thus, it is expected that in a warmer climate, the demand of water for irrigation is going to rise to alleviate moisture stress in soil or to meet the higher ET.

Change in temperature due to climate also influences the nutrient-use efficiency. In a simulation study, Jalota, Kaur, Kaur, and Vashisht (2013) pointed out that compared with time segment (1989–2010), the respective increase in temperature by 2.4°C and 5.1°C and CO_2 by 124 and 302 ppm during 2021–50 and 2071–98 time segments would decrease N-recovery efficiency (RE, Eq. 3.6) of applied $120 \, kg N \, ha^{-1}$ by 3.1% and 11.2% in rice and −3.1% and 4.1% in wheat:

$$RE(\%) = \frac{\text{Uptake in N fertilized treatment} - \text{Uptake in zero N treatment}}{\text{Quantity of N applied in fertilized treatment}}$$
$$\times \, 100 \qquad\qquad (3.6)$$

3.5 EFFECT OF RISING AIR TEMPERATURE ON PLANTS

Unlike CO_2, increasing temperature has a pessimistic effect on crop production. Temperature influences respiration and metabolism that affect the growth rate and production of assimilates in plants. Physiological responses to temperature changes in plants may occur at short or long time-scales. Short-term effects involve altered gene expression, such as heat-shock protein synthesis. Long-term responses include alterations in the rate of CO_2 assimilation, sucrose synthesis, and C and N partitioning within and between organs. Altered C availability brought about by these events affects uptake, transport, and assimilation of nutrients, resulting in changes in growth rates and yield. Temperature responses for specific physiological processes do not always relate directly to growth, because the growth is an integration of the effects of temperature on total metabolism. However, it is important to understand the effect of temperature on physiological processes like photosynthesis, respiration, transpiration, phenological developments in plants, and consequently biomass production and yield.

3.5.1 Photosynthesis

Response of temperature to photosynthesis is sigmoid. Net photosynthesis is less at lower and higher temperature than the optimum. Optimum temperature for photosynthesis is different in different crops, that is, 25°C in wheat, 21–27°C in cotton, 20–37°C in rice, 32°C in maize, and 34°C in soybean. Warmer environments accelerate leaf maturation and senescence, thus decreasing the duration of active photosynthesis. Photosynthetic efficiency decreases with temperature above 35–40°C. The optimum temperature for photosynthesis varies with CO_2 concentration. It (optimum temperature) is higher at elevated CO_2 level, because temperature controls the kinetic parameters of Rubisco and relative solubility of CO_2 and O_2, which means that photorespiration increases with temperature. In C4 plants, photosynthesis is tolerant to temperature than C3 due to the absence of photorespiration, which rapidly increases with temperature. In some crops like cotton, in which young population of leaves continuously being added to the top of the canopy, photosynthetic rate increases even in high-temperature environment.

3.5.2 Transpiration

An increase in air temperature increases directly the canopy temperature, leaf vapor pressure, and transpiration. The driving force in transpiration process is the gradient in vapor pressure concentrations between air spaces within the leaf and the surrounding air (Eq. 3.3). The effect of temperature on transpiration is mediated primarily through its effect on saturation of vapor pressure deficit of the air. Increased air temperature generally results in an increase in vapor pressure deficit because the amount of water needs to saturate a body of air rises exponentially with temperature. A formulation of saturated vapor pressure (e_{sat}) is given by Eq. (3.7):

$$e_{sat} = e_0 \exp \left[{(T-273)} \Big/ {(T-35)} \right] \tag{3.7}$$

where T is temperature in °K and e_0 is the saturation vapor pressure (0.61 kPa) at 273°K (0°C). The increased vapor pressure deficit in response to increased temperature increases the potential for transpiration and decreases water-use efficiency. Increased temperature transpiration is increased in both C3 and C4 plants, but the increase in transpiration per degree is less in C3 plants at higher temperature. For example, in soybean crop, evapotranspiration was 20% greater at 31°C and 30% at 35°C than 28°C. The less change in transpiration per degree increase in temperature over the range of 28–35°C than 28–31°C is due to evaporative cooling of leaves (Jones, Allen, & Jones, 1985). Air temperature also affects the transpiration rate depending upon the leaf area because transpiration increased linearly with leaf area. There were almost double transpiration rates of cotton plants at 36°C than those at 26°C (Reddy, Reddy, & Hodes, 1998).

3.5.3 Plant Development

Temperature affects plant development throughout their life cycle at all stages, that is, germination, emergence, vegetative growth, flowering, and grain fill. For the development of each crop, there are three cardinal temperatures, that is, base (T_b), below which development ceases; optimum (T_o); and ceiling (T_c), above which development ceases. Development events in plants occur faster at an optimum temperature provided that they have adequate levels of CO_2, water, sunlight, and plant nutrients. All these three temperatures are higher in C4 than many C3 plants (Table 3.2).

It means that response rate for a higher temperature is greater in C4 plants than C3. In most of the cases, phenological development is linked to the product of time and mean temperature above the base temperature for growth and development. Assuming that development increase is linear from baseline to ceiling temperatures, the rate of development is assessed in thermal time or growing degree days (GDD) with units °C day, as Eq. (3.8):

$$GDD = \sum_{i=1}^{N} (T_i - T_b) \tag{3.8}$$

where i is the ith day from sowing, T_i is the mean temperature for that day, T_b is base temperature, and N is the number of days in growing season. In case for a day minimum temperature is less than the base, then base temperature is taken as minimum for that day to calculate the GDD. Growing degree days needed to attain emergence, flowering, grain filling, and maturity in some crops are given in Table 3.3. However, these values can

Table 3.2 Cardinal Temperatures (°C) During Different Stages of Some Important Crops

Crop	Base/Critically Low Temperature			Optimum Temperature			Ceiling Temperature		
	G	V	F	G	V	F	G	V	F
Wheat	3–4.5	5	10–12	25	15–20	14–15	30–32	30–35	34
Rice	10–12	16	15	30–32	20–32	27–30	36–38	33	35–36
Soybean	9–13	18	13	15–32	26–32	27	40	35	39–40
Maize	5–10	—	—	32–35	32	—	40–44	40	35
Sorghum	8–10	10	15	21–38	26–29	25–28	40–48	—	35
Barley	3–4.5	—	—	20–21	15	—	38–40	—	—
Cotton	14	20	—	18–30	10	27–32	40–44	39	38

Table 3.3 Growing Degree Days for Growth Stages in Crops

Crop	Emergence	Flowering	Grain Filling	Physiological Maturity
Wheat	125–160	807–901	1068–1174	1538–1665
Chick pea	—	823–910	1037–1133	1394–1505
Mustard	108–136	650–718	868–945	1231–1322
Lentil	—	862–982	1028–1158	1673–1806
Peas	—	305–1451	862–982	1305–1451
Barley	109–145	738–936	927–1145	1269–1522
Oats	—	760–947	1019–1229	1483–1738
Sunflower	138–191	1081–1232	1255–1417	1780–1972
[a]Transplanted rice	—	1011–1124	1127–1214	1366–1500
[b]Rain-fed maize	100	1100	1250	1500

[a]Jalota et al. (2009).
[b]Jalota et al. (2010).
Modified from Miller, P., Lanier, W., & Brandt, S. (2001). Using growing degree days to predict plant stages. Montana State University Extension Service MT200103 AG 7/2001.

be different depending upon the soil type and water and crop management practices. Soils with less water retention, less water supply from rain or irrigation, low soil N, and more population of plants per unit area may have lesser degree days to complete a particular growth stage. Temperature exerts major control on growth at the level of enzyme activity, protein synthesis, and cell division. An increase in temperature accelerates crop respiration rate, shortens crop duration, or attains early maturity. Shortening of crop duration lessens the time to capture light, water, and nutrients during vegetative phase and reduces pollen production and their viability, the number and size of grain formation during the reproductive stage, and ultimately crop yield. In addition to these direct physiological effects, higher temperature has indirect effect through increased ET rate and hence water stress.

The magnitude of shortening of crop duration with increased temperature in some important crops is given in Table 3.4. Not only the total duration of the crop is decreased with an increase in temperature, but also individual growth stage is affected depending upon the magnitude of the increased CO_2 and temperature. There could be an increase in duration of a particular stage with increased temperature in combination with increased CO_2 under winter season crops during mid-century (MC, 2021–50) and end century (EC, 2071–98), shown in Table 3.5.

Table 3.4 Percent Decrease in Crop Duration in Different Crops With an Increase in Temperature

Reference	Crop	Experimental Condition/ Model used	Increase in Temperature (°C)	Shortening of Crop Period (Days)
Experimental				
Xiao et al. (2007)	Wheat	Green house	2	
Asseng et al. (2004)	Wheat	APSIM	4.2	13
Batts et al. (1997)	Wheat	TGT	4	42
Reddy et al. (1992)	Cotton	CERES-cotton	5	10
Simulations				
Kaur and Hundal (2010)	Wheat	CERES-wheat	1, 2, 3	6, 12, 17
Jalota, Kaur, Kaur, et al. (2013)	Wheat	CropSyst	2.4, 5.1	9, 23
Vashisht et al. (2013)	Wheat	CERES-wheat	4.4	8
Kaur and Hundal (2010)	Rice	CERES-rice	1, 2, 3	1, 1, 5
Jalota, Kaur, Kaur, et al. (2013)	Rice	CropSyst	2.4, 5.1	18, 25
Kaur and Hundal (2010)	Soybean	CERES-soybean	1, 2, 3	1, 2, 2
Kaur and Hundal (2010)	Groundnut		1, 2, 3	2, 5, 9
Kaur and Hundal (2010)	Chickpea	CERES-chickpea	1, 2, 3	8,16, 24
Jalota et al. (2009)	Rice-wheat	CropSyst	4.6	27
	Maize-wheat	CropSyst	4.6	32
	Cotton-wheat	CropSyst	4.6	53

Table 3.5 Duration of Different Phonological Stages of Rice and Wheat Crops in Different Time Segments

	Rice				Wheat			
	F[a]	F-GF	GF-M	Total	F	F-GF	GF-M	Total
Time Segment	Duration (days)							
Present (1989–2010)	65	8	32	105	115	10	37	162
Mid century (2021–50)	56	7	24	87	95	13	45	153
End century (2071–98)	52	6	22	80	82	12	45	139

[a]F, FG, and M stand for flowering, grain filling, and maturity stages, respectively.

3.5.4 Biomass and Yield

Different crop species have different optimum T_{max} and T_{min} ranges for normal physiological processes to go on. In rice, wheat, maize, and cotton crop, the range of optimum T_{max} is 25–37°C, 20–25°C, 25–32°C, and 27–32°C, respectively. The corresponding ranges of T_{min} are 24.3–25.5°C, 8.6–10.9°C, 24.6–26.2°C, and 21.9–23.9°C. Plant growth is faster at optimum temperature, which is generally higher for vegetative than reproductive development. Temperature affects the crop yields by affecting the crop-specific yield attributes, for example, duration of anthesis and grain filling in wheat (Aggarwal & Kalra, 1994), cotton sterility and boll retention in cotton (Sankaranaryanan, Praharaj, Nalayani, Bandyopadhyay, & Gopalakrishanan, 2010), spikelet sterility in rice (Ohe, Saitoh, & Kuroda, 2007), and kernel weight in maize (Singletary, Banisadr, & Keeling, 1994). In general, yields of agronomic crops are reduced with increased temperature from normal and vice versa.

For example, a temperature increase of 1–2°C compared with normal conditions decreased simulated grain yield by 2%–5% in rice and 14%–23% in wheat, while with the same decrease in temperature, an increase in yield was 8%–15% in rice and 7% in wheat (Hundal & Kaur, 2007). An increase in the temperature in low-temperature regions can reduce low-temperature limitations like frost risk and can cause an increase in crop yield by accelerating the crop growth provided that soil moisture is not limiting. In Chitral district of Pakistan, wheat yield did increase by 14%–23% with the increased temperature of 1.5–3.0°C (Hussain & Mudassar, 2007). Data from nine different sites (258 yield observations) revealed 6% decline in wheat yield per degree Celsius increase in temperature during growing season, from germination to maturity (Bender et al., 1999). In soybean, seed response peaked between 32/22°C and 36/26°C (T_{max}/T_{min}) and declined rapidly at higher temperature (Pan, 1996). Cotton plants grew faster at 30/22°C than at either higher or lower temperatures. However, the plants at 35/27°C had more boll weight than those grown at 30/22°C, and they were more advanced in fruiting structure formation (Reddy et al., 1992). In potato, the optimal temperature range for tuber growth is between 16°C and 22°C (Kooman, 1995). Daily temperatures that are outside this optimal range result in reduced tuber growth and lower assimilation to tubers, HI, and tuber yield.

It is not only the changed temperature during the whole growth period deciding crop production but also the stage at which temperature changes that is important. Each crop in its life cycle requires an optimum temperature at different phenological stages (Table 3.2), and it is the vulnerability of crop growth stage to high/low temperature under changed climate that determines the ultimate yield. For example, in wheat, the most sensitive stage is grain fill, and the optimum temperature for that stage is 25–35°C. During grain fill, an increase in temperature by 1°C typically shortens its duration and reduces HI and grain yield proportionally (Batts et al., 1997; Mitchell et al., 1995; Pushpalatha, Sharma-natu, & Ghildiyal, 2008; Wheeler, Craufurd, Ellis, Porter, & Vara Prasad, 2000). During grain fill, it is the increase in T_{max} in wheat (Aggarwal & Kalra, 1994; Jalota, Kaur, Kaur, et al., 2013) and T_{min} in rice (Peng et al., 2004; Prasad, Boote, Allen, Sheehy, & Thomas, 2006; Welch et al., 2010) that cause detrimental effect on grain yield in tropical and subtropical Asia. In rice, optimum temperature for grain fill is 20–25°C (Yoshida, Satake, & Mackill, 1981), which is lower than other development stages. Temperature below 20°C causes a failure in pollen development at microspore stage (Stake & Hayase, 1970) or sustains injury at flowering (Abe, 1969). In general temperature, >25°C results in poor grain fill as it reduces grain-fill duration, which in turn results in reduced grain yield. Summarizing data from several experiments, Baker, Allen, and Boote (1995) concluded that grain yield of rice declines by about 10% for every 1°C rise in daily mean temperature above 26°C to zero yield near 36°C, which is due to spikelet infertility (Matsui, Omasa, & Horie, 1997). A decline in rice yield at temperature >34°C from flowering to anthesis period has also been observed in Punjab state of India by Chahal, Sood, Jalota, Choudhury, and Sharma (2007). In cotton, the time period in plant cycle that is vulnerable to high temperature is short, that is, just before, during, and after the flowering. Boll growth is sensitive to temperature, and the bolls are usually abscised immediately after flowering when exposed to high temperature of >28°C (Reddy, Reddy, Acock, & Trent, 1994). Young bolls are injured by exposure to only a few hours of such high temperature. The upper limit for fruit survival is ~32°C. In groundnut (*Arachis hypogaea* L.), fruit setting stage is sensitive to high temperature. Fruit set got reduced by 7% per degree increase in temperature when temperature exceeded critical limit of 38°C, and there was no fruit set when temperature reached to 43°C (Vara Prasad, Craufurd, Summerfield, & Wheeler, 2000). The adverse effect of temperature is more in drought conditions because an increase in temperature increases the evaporative demand and exacerbates plant stress. This effect is more in fine- than coarse-textured soils (Jalota, Kaur, Kaur, et al., 2013; Ludwig & Asseng, 2006). In general, the relationship between temperature and yield is negative, which is attributed to shortening of the optimal growth period. The magnitude of reduction in yield with increased temperature in some important crops is given in Table 3.6. Using data of wheat crop from Table 3.6, a response function (polynomial) of wheat yield reduction to increased temperature has been

Table 3.6 Percent Reduction in Yield of Different Crops With Change in Temperature

Reference	Crop	Method/ Model Used	Increase in Temperature (°C)	Reduction in Yield (%)
Experimental				
Xiao et al. (2007)	Wheat	Green house study	2	9.4
Asseng et al. (2004)	Wheat	APSIM	3	13
Batts et al. (1997)	Wheat	TGT	4	42
Li et al. (2005)	Wheat	Field	1.8	21
Li et al. (2005)	Wheat	Field	0.8	11.6
Chowdhury and Wardlaw (1978)	Wheat		1	4
Bender et al. (1999)	Wheat		1	6
Bender et al. (1999)	Wheat		1	5.4
Peng et al. (2004)	Rice	Field	1 (T_{min})	10
Baker et al. (1995)	Rice		1	10
Muchow et al. (1990)	Maize		1	2.5
Runge (1968)	Maize		1	3.2
Lobell and Field (2007)	Maize		1	8
Reddy et al. (1992)	Cotton	CERES-cotton	1	3.5
Pettigrew (2008)	Cotton lint		1	10
Chowdhury and Wardlaw (1978)	Sorghum		1	8.4
Lobell and Field (2007)	Soybean		1	1.3
Boote et al. (1998, 1997)	Soybean		1	3
Prasad et al. (2003)	Peanut		1	4.5
Prasad et al. (2002)	Beans		1	7.2
Simulations				
Kaur and Hundal (2006)	Wheat	CERES-wheat	1, 2, 3	9.9, 18, 27
Pandey et al. (2007)	Wheat	CERES-wheat	1	8
Lal et al. (1998)	Wheat	CERES-wheat	4	54
Mitchell et al. (1995)	Wheat	ARCHWHEAT	4% higher than base	35
Vashisht et al. (2013)	Wheat	CERES-wheat	4.4	61
Kaur et al. (2012)	Wheat	CropSyst	1, 3	0.4, 14.9
Hundal and Kaur (1996)	Wheat	PlantGrow	1, 2, 3	8.1, 18.7, 25.7
Hundal and Kaur (1996)	Rice	PlantGrow	1, 2, 3	5.4, 7.4, 25.1
Kaur and Hundal (2006)	Rice	CERES-rice	1, 2, 3	2.8, 9.6, 10.1
Lal et al. (1998)	Rice	CERES-rice	4	41
Hundal and Kaur (1996)	Maize	PlantGrow	1, 2, 3	10.4, 14.6, 21.4
Kaur et al. (2012)	Maize	CropSyst	1, 3	13.7, 37.1
Kaur and Hundal (2006)	Soybean	CERES-soybean	1, 2, 3	−2.4, −2.4, 5.6
Kaur and Hundal (2006)	Groundnut	CERES-groundnut	1, 2, 3	4.5, 10.6, 13.1
Vashisht et al. (2015)	Potato	SUBSTOR-potato	6, 2.1	19, 29
Hundal and Kaur (1996)		PlantGrow	1, 2, 3	8.7, 23.2, 36.2

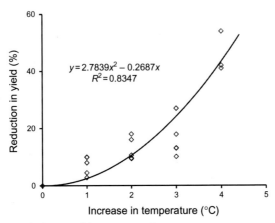

$$y = 2.7839x^2 - 0.2687x$$
$$R^2 = 0.8347$$

Fig. 3.3 Response of wheat yield to increased temperature.

worked out, which shows 3% reduction in wheat yield by 1°C increase in temperature than the near optimum, which rose to 12% by 2°C, and thereafter, the yield reduced abruptly (Fig. 3.3). The nonlinear response may be because of varying experimental conditions of climate, soil, and management practices, interactions of water and heat stress during hot days, and most sensitivity of the reproductive period to temperature for flower fertility. Increasing temperature has also an impact on grain quality of crops. Quality of marketable yield in different crops is assessed based on different criteria. In rice, grain quality indicators are milling efficiency (head rice yield), shape and appearance (grain length before and after cooking, grain width, and chalkiness), cooking and edibility characteristics (amylase content of the endosperm, gelatinization temperature, and aroma), and nutritional quality (protein, oil, and micronutrient content) (Resurreccion, Hara, Juliano, & Yoshida, 1977). High temperature lowers amylose content and results in a higher palatability (Huang & Lur, 2000), but does not affect protein contents (Cooper, Siebenmorgen, & Counce, 2008). High–night-temperature condition does not decrease grain quality (Morita, Shiratsuchi, Takanashi, & Fujita, 2004). In wheat, the quality indicators are baking quality, dough strength (resistance to deformation), and loaf score (Randall & Moss, 1990). High-temperature regime after anthesis affects accumulation of protein (arabinoxylans) during grain fill (Hurkman, Vensel, Tanaka, Whitehand, & Altenbach, 2009) and increases the grain arabinoxylan concentration (Zhang, liu, Chang, & Anyia, 2010). Water-soluble arabinoxylans are important in baking quality of wheat because of their effect on the viscosity and dough. Dough strength increased with protein concentration and tends to decrease at higher temperature. Higher mean temperature is positively correlated to gluten quality (Moldestad, Fergestad, Hoel, & Skjelvag, 2011). Wheat grains developed at high temperatures produce poorer quality flour for making bread. In maize, high temperature reduces starch, protein, and oil contents of the kernel. Amylopectin chain gets lengthened with increased temperature.

At the molecular level, high temperatures could change gene expressions for amylose and storage protein accumulation and further affect taste quality (Lin et al., 2010; Lin, Chang, Tsai, & Lur, 2005; Yamakawa, Hirose, Kuroda, & Yamaguchi, 2007). In cotton, the quality indicator is fiber strength, which is more in warmer environment (Pettigrew, 2008).

In nutshell, increased temperatures hasten nutrient mineralization in soil; decrease fertilizer-use efficiency; increase crop respiration rates; reduce crop duration, the number of grains formed, and crop yield; and deteriorate grain quality.

3.5.5 Effect of Extreme Climatic Conditions

The climate extremes may or may not affect the biomass, but affect yield definitely. Impact of climate extremes on crops is decided by the climatic parameter and the stage of crop development. Extreme cold, frost chilling, hot temperatures, and P_{cp} of high intensity at reproductive stage have detrimental effect on crop yields. Temperature extremes during the reproductive stage of development can produce some of the largest impacts on crop production because of the effect on pollen viability, fertilization, and grain or fruit formation. In wheat, chilling ($<5°C$) and hot ($>30°C$) temperatures at anthesis can damage pollen formation, which in turn reduces seed set and can decrease yield (Batts et al., 1997; Tashiro & Wardlaw, 1990). In soybean, pod development and seed filling are most sensitive phases of crop development. High temperature at time of flowering or early seed filling and water deficit during reproductive stage reduce photosynthesis and seed yield of soybean (Ferris, Wheeler, Hadley, & Ellis, 1998). Though extreme heat decreases yield, reduction in yield by droughts is more due to reduction in both harvested area and yields as a result of crop failure. Excess precipitation causes aeration problem in soil and physical damage to crops like washing away of pollens, lodging and rotting. A number of extreme, record-breaking events have been experienced at global level that has caused crop and human loss. For example, the year 2011 was the hottest and driest spring since 1880 in Western Europe (France) that caused 12% grain harvest loss (IPCC, 2011). Recently, in an analysis of 46 years (1964–2007), Lesk, Rowhan, and Ramankutty (2016) concluded that droughts have decreased production by 7% at global level and 8%–11% in developed countries than in developing ones. In Maharashtra state of India, drought conditions are occurring for the last 3 years, that is, from 2013 to 2016. Two hundred human deaths were recorded in 2009 due to heat wave in Victoria, Australia. There was a hottest summer since 1891 during the year 2007 in Southern Europe (Greece) and extreme July heat wave during 2011 in Texas and Oklahoma (the United States). Damage of crops and human lives caused by flooding are the evidences of weather extremes in England and Wales during 2007, Pakistan and eastern Australia during 2010, and Punjab state of India during 1988. Hurricanes during the years of 2005 and 2011 in the United States also caused damage to crops and humans.

3.6 EFFECT OF PRECIPITATION

Altered P_{cp} patterns may impact the soil by increasing erosion risk, runoff, and nutrient losses. Increased erosion risk stems from the increased events of heavy P_{cp}, especially in heavy textured soils because of the inability of the soil to maintain the infiltration enough to absorb high-intensity P_{cp}. As a result, rain water is lost as runoff, which is detrimental to soil health and crop productivity. Runoff causes soil erosion and washes away stored C, nutrients, and applied agricultural chemicals. In coarse-textured soils, nutrients and agricultural chemicals are leached into ground and surface waters. The soils with low organic matter are most vulnerable to erosion. Sometimes severe erosion events can lead to unintentional landscape conversions. Low P_{cp} accompanied by high temperature results in droughts that inhibit nutrient cycling and encourage an upward movement of soil water by evaporation that causes salinization.

The effect of P_{cp} on plants is mediated through water retained in the soil. The proper soil moisture regimes make better availability of the nutrients and promote root growth by providing optimum physical condition of less soil strength. As plant water status is directly related to the soil water, therefore, physiological processes like photosynthesis, evapotranspiration, and crop development during different growth stages also increase with the increased soil water. On the other hand, if soil water availability is decreased or ET is increased at the most sensitive stage of the crop, yield declines drastically. The water-sensitive growth stages for different crops have already been defined. For example, pod setting and grain filing in chickpea, anthesis to grain filling in wheat, silking and tasseling in maize, boll formation in cotton, pod formation and pod filling in soybean, and anthesis to achene filling in sunflower. With increased water supply from P_{cp} or applied as irrigation, physiological processes become better and increase crop yield ultimately. Small and frequent rain showers retain soil moisture for longer period in the crop root zone and are beneficial for plant growth. High P_{cp} has some negative effects too. It increases the potential for increasing susceptibility to root diseases, delays harvest, and deteriorates the quality of many crops as a result of increased disease infestation on grains. More intense storms associated with turbulence and wind gusts increase potential for lodging. Precipitation during reproductive period (flowering to anthesis) in rice may affect pollination and fertilization and ultimately yield by damaging/washing away the pollens (Chahal et al., 2007).

In the existing climate records and projections, there is more uncertainty and variability in all the characters of P_{cp} like total, frequency, and intensity at spatial and temporal scales. More variability in these characters causes more variability in available soil water and crop yields. This variability is more in nonirrigated crops/cropping system or rain-fed agriculture than irrigated ones. The higher magnitude of variation in nonirrigated crops/cropping system though can be reduced with irrigation, but cannot be eliminated. Crop yields are often having negative correlation with the meteorological droughts, but quality is certainly improved. In drought conditions, concentration of arabinoxylans in wheat is increased (Coles et al., 1997). Drought increases grain hardness in wheat. However, vitreosity and protein content are poorly related to drought severity (Weightman et al., 2008).

3.7 INTERACTIVE EFFECT OF ELEVATED CARBON DIOXIDE AND TEMPERATURE

Global surface temperature may rise within the range of 0.4–2.6°C in 2046–65 and 0.3–4.8°C in 2081–2100 relative to the reference period of 1986–2005 (IPCC, 2014). Level of CO_2 at the end of 21st century may reach 883 ppm (http://co2now.org). Both the parameters CO_2 and temperature are showing an increase and would reach the higher value than these at present. As discussed earlier, the increased CO_2 increases crop yields, and increased temperature beyond optimum decreases crop yields. Here, one question that arises is that how much increase in one (favorable) variable will offset the adverse effect of the other (detrimental) variable and vice versa? While studying the effects of CO_2 and temperature on different crop species, the studies of Baker et al. (1992) on rice; Batts et al. (1997) on wheat; Reddy, Hodges, and Kimball (2000) on cotton; and Vashisht et al. (2015) and Wolf and Van Oijen (2003) on potato established that at optimum and little higher degree of temperature, CO_2 works opposite the negative effect of high temperature and compensates for the decrease in crop yield. Das et al. (2007) stated that positive effect of doubled CO_2 in rice persists only up to $+1°C$ warmer climate than the normal. But beyond a certain value, negative effect of temperature overweighs the positive effect of CO_2, and there is a decrease in crop yield. So, net positive effect of CO_2 may occur in some years when the temperature increase is less and not in others (Hakala, 1998).

A number of studies, mostly simulations, have been made to quantify the temperature increase that can nullify the beneficial effect of increased CO_2 levels and vice versa. The values of such temperature or CO_2 reported in different studies are quite variable. For example, a study by Hundal and Kaur (2007) points out that rice crop was unable to offset the adverse effect of $1°C$ increase in temperature from normal at CO_2 concentration of 500 ppm. The beneficial effect of increased CO_2 from 330 to 660 ppm was nullified by a rise in temperature of $3°C$ in rice and soybean and $2°C$ in wheat (Lal et al., 1999). Similarly, the affirmative effect of increased CO_2 concentration from 350 to 700 ppm was nullified by the increased temperature of 1 (Kaur et al., 2012) and $0.6°C$ (Sahoo, 1999) in maize and $0.9°C$ in sorghum (Chaterjee, 1998). Yield loss of rice by $2°C$ rise in temperature was compensated by 425 ppm CO_2 (Saseendran et al., 2000). All these and other studies of similar type have limitations of variations in starting point, upper point of CO_2 concentrations and temperatures, and the models being used for evaluation of their interactive effect. Such limitations can be removed by developing a model by which the interaction can be interpolated within a wide range of temperature and CO_2 concentration. Jalota et al. (2009) and Kaur et al. (2012) developed multiplicative models (Eq. 3.9) from the data generated from simulation studies on relative yield as a function of relative CO_2, T_{max}, and T_{min}. Using the model, they have assessed interactive effect of CO_2 and temperature on rice, maize, and cotton crops and wheat equivalent to yield of rice-wheat, maize-wheat, and cotton-wheat cropping systems under varying temperature and CO_2 levels:

$$\left(Y/Y_a \right) = \alpha \prod \left(CO_2/CO_{2e} \right)^{\gamma 1} \left(T_{max}/T_{maxe} \right)^{\gamma 2} \left(T_{min}/T_{mine} \right)^{\gamma 3} \qquad (3.9)$$

where Y is crop yield (kg ha^{-1}); Y_e is existing crop yield (kg ha^{-1}); CO_2 is elevated concentration (ppm); CO_{2e} is existing concentration (ppm); T_{max} is maximum temperature (°C) and T_{maxe} is existing maximum temperature (°C);

Table 3.7 Intercept, Sensitivity Factors, and Coefficient of Correlation of the Equations Fitted to Different Crops and Cropping Systems

Crop	[a]α	$\gamma 1$	$\gamma 2$	$\gamma 3$	R^2
Rice	1.017	0.069	−1.442	−0.774	0.97
Wheat	1.002	0.163	−1.408	−0.425	0.99
Rice–wheat	1.011	0.105	−1.428	−0.638	0.99
Maize	1.006	0.092	−2.987	−1.459	0.95
Wheat	0.998	0.117	−0.568	−0.088	0.97
Maize–wheat	1.013	0.106	−1.603	−0.703	0.97
Cotton	1.052	0.077	−4.534	−1.806	0.94
Wheat	1.004	0.053	−0.578	0.128	0.98
Cotton–wheat	1.026	0.066	−2.703	−0.966	0.96

[a]α is intercept; R^2 is coefficient of determination; and $\gamma 1$, $\gamma 2$, and $\gamma 3$ are sensitivity factors for CO_2, maximum temperature, and minimum temperature, respectively.

T_{min} is minimum temperature (°C) and T_{mine} is existing minimum temperature (°C); $\prod$ is multiplication; and $\gamma 1$, $\gamma 2$, and $\gamma 3$ are the sensitivity coefficients for CO_2, T_{max}, and T_{min}, respectively. The values of these coefficients are negative for temperature and positive for CO_2 (Table 3.7). The coefficients of determination are very high indicating that yield variation can be explained by these factors.

More negative values of coefficient for T_{max} indicate that yield reduction by an increase in T_{max} is more than by T_{min}. The interaction shows that the beneficial effect of doubled CO_2 level (760 ppm to the base 380 ppm) is leveled off with increased T_{max} by 1.1°C of the existing temperature in rice crop. The crop productivity is also influenced by interactions of temperature × precipitation. In an analysis of data from 1976 to 2008 for Wisconsin, Kucharik and Serbin (2008) and Serbin and Kucharik (2009) found that temperature effect could be offset by increases in P_{cp}. In Indian conditions, 5%–10% increase in P_{cp} has been found enough to absorb the adverse effects of rise in temperature by 1–2°C (Abrol, Bagga, Chakravarty, & Wattal, 1991).

In rain-fed crops, application of irrigation can increase the positive effect of CO_2 by depressing the yield reduction effect of temperature (Xiao et al., 2005).

3.8 PARTITIONING OF CARBON DIOXIDE AND TEMPERATURE EFFECTS

Under field conditions, it is intricate to compute the contribution of affirmative effect of CO_2 in lowering down the negative effect of yield reduction by increased temperature. Simulations can abet to study such effects. Jalota, Kaur, Ray, et al. (2013) quantified such effects in rice and wheat crops by running the CropSyst model with and without increased CO_2 for the years 2020, 2050, and 2080. They concluded that in the years 2020, 2050, and 2080 compared with the present time (1989–2010), increased CO_2 concentrations of 40, 70, and 300 ppm (from 380 ppm) would be able to lower the yield reduction by increased temperature of 1.4°C, 2.6°C, and 3.8°C to the tune of 4.7%, 9.7%, and 12.5% in wheat and 4.4%, 9.4%, and 12.9% in rice crops, respectively. The data compiled by Hatfield et al. (2011) in the review also portray that 60 ppm increase in CO_2 (from 380 ppm) could lower down the reduction in yield through 0.8°C increase in temperature (from 19.5°C) by 6.8% in wheat, 6.4% in rice, 9.2% in cotton, 6.7% in peanut, 6.1% in beans, 7.4% in soybean, and 1% in maize. The benefit of increasing CO_2 also depends upon the enormity of the increased temperature. For instance, the benefit of increasing CO_2 by 90 ppm (from 360 ppm) lowered down the yield reduction effect in rain-fed spring wheat by 16.9% with increased temperatures of 0.8°C (from 14.3°C) and by 14.9% with 1.8°C increase in temperature (Xiao et al., 2005).

3.9 INTERACTIVE EFFECT OF CARBON DIOXIDE AND NITROGEN

Elevated CO_2 alters the internal balance between C and N in most of the plants by increasing C and decreasing N. Under elevated CO_2, C is increased because of increased photosynthetic activity. Nitrogen is decreased due to the dilution of N from increased carbohydrate concentrations and decreased uptake of N and other minerals from the soil with decreased stomatal conductance. Under elevated CO_2 (from 359 to 534 ppm), there was 13% decrease in leaf N concentrations in plant tissues (Ainsworth & Long, 2005). Elevated CO_2 decreases plant N in different plant species following the order of C4 > C3 nonlegumes > C3 legumes plants. Increased CO_2 increases sink capacity of the growing shoot, root, and other vegetative plant parts and of grain during grain-filling stage of the plant for N source. The source-sink relations between growth and N supply and correlation of the sink strength with the N availability to wheat plant have been established by Rogers, Milham, Gillings, and Conroy (1996). At high CO_2 (900 ppm) and low N treatments, there was no gain in the shoot growth, but when N supply was increased (to $67\,mg\,N\,kg^{-1}\,soil\,week^{-1}$, keeping CO_2 900 ppm), the growth became double than the previously observed at low N treatments. Similar trends of growth in response to

N application were observed in sensitivity analysis for spring wheat by Asseng et al. (2004). At low N ($100\,kg\,N\,ha^{-1}$) and adequate irrigation water supplies, yield was less by 8% at elevated CO_2 of 500 ppm than at ambient level of CO_2 (368 ppm), but when N was increased from 100 to $400\,kg\,N\,ha^{-1}$, it gave maximum yield output of $8.4\,t\,ha^{-1}$. Thus, CO_2 and N interaction clearly indicates that under elevated CO_2 conditions, N requirement of C3 plants may increase in the future and not in C4 plants (Hocking & Meyer, 1991). However, nitrogen–use efficiency (NUE) would be higher in C4 plants than that of C3.

3.10 CROP RESPONSE TO CLIMATIC VARIATION

Climate change and variability are considered as one of the main environmental problems of the 21st century, and the climate has already experienced a shift toward the increased variability in some climate parameters. Variability in the data is generally expressed as Eq. (3.10):

$$\text{Variance} = \frac{\text{Standard deviation}}{\text{Mean}} \qquad (3.10)$$

The climatic variability creates meteorological conditions that deviate significantly from mean. Significant changes in the mean and variance make the weather extreme (IPCC, 2007; Porter & Gawith, 1999). The changes may be in (i) the mean weather, which is likely to increase, that is, annual mean temperature and/or P_{cp}; (ii) the distribution of weather, such that there are more frequent extreme weather events such as physiologically damaging temperatures and P_{cp} or longer periods of drought; and (iii) a combination of changes to the mean and its variability. In agroecosystems, inter- and intra-seasonal variability at spatial and temporal scales is a key concern, but the important question for agriculture is how any change in the variability of weather at a particular stage of the crop is reflected in changes in the variability of crop production? Mean and variability of temperature affect crop growth, development, and yield by influencing the processes of photosynthesis, respiration, and phenology. Temperature variability is not believed to be affecting yield if change in yield with change in temperature is proportional. There are certain optimum temperatures at which plants give potential yield and variation from that causes variability in yield. For example, a little (1°C) above that temperature may not have significant effect on potential yield of wheat, but a further increase (to 2°C) reduces the crop yield (Aggarwal & Sinha, 1993; Fig. 3.3). The variation in yield reduction is region-specific; for example, with 2°C increase in temperature, the decrease in potential yield of wheat was small (1.5%–5.8%) in subtropical, but in tropical regions, the decrease was 17%–18% (Aggarwal & Kalra, 1994). With increased temperature, wheat yield decreases at lower latitude but increases at higher. In India, the yield increased at latitude greater than 27°N (Mall, Singh, Gupta, Srinivasan, & Rathore, 2006). Generally, crops exposed to higher temperatures at the developmental stage experience effect of damage, but experimental studies of the effects of temperature variability on crop productivity are rare. This is mainly because of the difficulties in establishing and maintaining a temperature regime where a mean climatic value can be held constant between treatments that vary the amplitude of temperature. Like temperature, the variability of P_{cp} both within and between season and humidity regimes is of greatest challenge in semiarid conditions, as these have substantial effect on crop growth, pest incidence, and crop productivity. Response to rainy season is affected by the amount of P_{cp} and the subsequent amount of water stored in the soil for the crop. In addition to temperature and P_{cp}, light variability also causes variation in yield of crops; however, the independent effects of temperature and P_{cp} were three to four times larger than caused by variation in incident light (Monteith, 1981). The effect of climate variation is less in simulation studies than actual because effects of yield-reducing factors (weeds, pests, diseases, and pollutants) are not taken care in the former (Reidsma, Ewwert, Lansink, & Leemans, 2010). Furthermore, the impact of climate variability on crop yield is more at regional scale than the farm scale because of the dynamic complexity of the climate, soil, and plant (Subash & Ram Mohan, 2012).

Although extreme values, especially of temperature and P_{cp}, can have large effects on crop yields and their variability, less is known about changes in the variability of climate. However, simulation modeling of the effects of climatic variability has pointed to the general conclusion that increased annual variability in weather causes increased variation in yields. Porter and Semenov (2005) calculated that variation in temperature, radiation, and P_{cp} causes yield variation of 12% on fine-textured and 17% on coarse-textured soils in winter cereals. Under field conditions, the effect of climate variability is also compounded by effect of management interventions. In a field study of 6 years (from 2008–09 to 2013–14) on wheat crop at Ludhiana in central Punjab (India), Vashisht and Jalota (2014) delineated the contribution of variations in temperature and management interventions on yield variation. Out of the total variation of 12% in the wheat yield, 8% was caused by climate and 4% by management interventions such as date of planting, cultivars, and irrigation schedules. It was also pointed out that 8.2% variability in wheat yield at present (2008–14) caused by variability of 3.4% in T_{max} and 4.6% in T_{min} would be changed to 11.2% by variability of 5.5% in T_{max} and 3.8% in T_{min} in time slice of 2031–50. The effect of changed temperature on crop yield also varies with the crop growth period and sowing time. For example, in present climate scenario of Punjab state in northwest India, an increase in temperature during the fourth week of January to the first fortnight of February reduces the yield of early sown wheat (the fourth week of October) by 1% $°C^{-1}$; during the first to second fortnight of February affects the yield of normal sown (November 8th) by 1.9–3.3% $°C^{-1}$; and during the first fortnight of March to the first week of April of late sown (from November 25th to December 2nd) by 3.4–3.5% $°C^{-1}$ (Kaur & Hundal, 2010). It is expected that due to climate change in the future, these variations would increase with increased number of hot days. Under such situations, crop yield would be affected by the number of days in conjunction with the magnitude of increased temperature. Recently, Vashisht and Jalota (2014) developed relations between intraseasonal temperature and wheat yield, from the field data of 6 years, as weather descriptors (Table 3.8). The relations between yield and weather descriptors for Ludhiana location indicate that weather descriptors, namely, days <20 with $T_{max} \geq 27°C$, days <30 with $T_{min} \geq 11°C$, days <3 with $T_{min} \leq 3°C$, and days <1 with $T_{max} \geq 34°C$, prohibit a decline in wheat yield. However, these may vary with crops and locations (Bannayan & Sanjani, 2011; Porter & Gawith, 1999). Actually, more extreme heat (>34°C) exposure toward the end of the crop cycle slows photosynthesis and grain-filling rates (Ritchie & NeSmith, 1991; Tashiro & Wardlaw, 1989). More number of days with $\leq 3°C$ adds to lowering of yield by retarding physiological processes including photosynthesis, respiration, nutrient movements, and transpiration, which negatively affects plants growth (Gill & Tuteja, 2010).

Table 3.8 Wheat Yield as Influenced by the Number of Days of Different Temperatures

Year	Yield (kg ha^{-1})	T_{max} >27°C	T_{min} >11°C	T_{min} <3°C	T_{max} >34°C
2008–09	4176	41	54	0	1
2009–10	4377	35	50	3	11
2010–11	5020	34	45	4	0
2011–12	4249	32	46	15	4
2012–13	4064	32	52	3	1
2013–14	4914	18	32	3	1

Table 3.9 Equivalent Rice Yield in Present Time Slice and Percent Reduction in Mid and End Century at Different Locations

Location	Equivalent Rice Yield (Mg ha^{-1}) Present Time Slice (1989–2010)	Percent Reduction in Equivalent Rice Yield Mid Century (2021–50)	End Century (2071–98)
Amritsar	12.4	5.8	24.9
Jalandhar	11.8	4.7	11.8
Ludhiana	11.5	8.5	10.1
Patiala	11.8	1.8	23.6
Average	11.9	5.2	17.8
Variance (%)	2.9	53.3	43.9

Equivalent rice yield = yield of rice + (yield of wheat × price of wheat/price of rice).

So, the developed weather descriptors highlight the importance of duration of temperature more than the optimum or less than the base during crop season and at a particular stage that hampers the crop yield. Spatial variation in weather during the growing season and soil types at different locations also cause variation in yield. From the data of Jalota, Vashisht, Kaur, Kaur, and Kaur (2014), variation in the reduction of rice equivalent yield at four locations of central Indian Punjab was 53.3% during MC and 43.9% during EC (Table 3.9). This was caused by the variation of 10.1% in T_{max} and 11.7% in T_{min} during MC and 18.2% in T_{max} and 12.6% in T_{min} during EC along with variation in soil texture.

3.11 CROP RESPONSE TO CLIMATE CHANGE

The climate changes make their impact on crop productivity mainly through increased/decreased temperature, an increase/decrease in P_{cp}, and increased CO_2 and ensuing incidence of insect pest attacks and weed growth. The individual effect of CO_2, temperature, and P_{cp} on plant physiological processes, growth, and yield has already been discussed. But it is important to have an awareness of integrated effect of changed climatic parameters on crop productivity. Effect of climate change can be assessed from the past observations and the projected data. In the latter case, data of climate change scenarios (based on the driving force population, economy, technology, energy, land use, and agriculture) are used. From the climate change scenario, data on changes in temperature, P_{cp}, and CO_2 concentration are projected from multiple climate model assemblages (details given in Chapter 2). Using the projected data of climate change, yields of crops and water-balance components are assessed with crop models for different segments of the 21st century. At present, climate change is not a scientific prognosis, but is a reality. For example, Lobell, Schlenker, and Costa-Roberts (2011) have concluded that yields of maize (*Zea mays* L.) and wheat (*Triticum aestivum* L.) have already declined by 3.8% and 5.5%, respectively, compared with the yields without climate change in North America. In European countries, for two decades, there is a decline in growth trends of cereal yields, especially wheat (Olesen et al., 2011). There are trends of decline and abandonment of agriculture in some parts of Mediterranean and southeastern regions and an intensification of agriculture in Northern and Western Europe (Stoate et al., 2009). Roudier, Sultan, Quirion, and Berg (2011) assessed the potential impact of climate change on yield in West Africa and predicted 5% larger loss in Sudano-Sahelian than southern Guinean countries due to drier and warmer conditions.

In recent years, India has experienced frequent droughts, floods, and production losses. In northern part of Himalayas region, the apple belt has been shifted to higher elevations to have adequate chilling, essential for apple yields (Rana, Bhagat, Kalia, & Lal, 2010). Jones and Thornton (2003) analyzed the possible impacts of climate change on maize production in Africa and Latin America. They projected a yield decrease in 2055 owing to increased temperature and P_{cp}, less conducive for maize production. Using the regional climate model and the A2SERES scenario, Evans (2009) estimated that in North Africa, maize yield was projected to fall between 15% and 25% with 3°C rise in temperature. Reduced P_{cp} and increased temperature are expected to have negative effects on maize, wheat, and soybean in Brazil (De Siqueira, Farias, & Sans, 1994). Rosenzwig, Parry, Fischer, and Frohberg (1993) indicated that northeastern Brazil would suffer severely in the world due to climate change impact on food production. Rain-fed crops are likely to be benefited in some regions due to an increase in P_{cp}. In South Asia, rice production projected to reduce by 2%–10%; yields of irrigated maize, sugarcane, and sorghum are likely to decrease by 7%–25 % depending upon the scenario and time (Knox, Hess, Daccache, & PreezOrtola, 2011). Wheat yield in Southern Australia would decrease from 13.5% to 32% in 2080 with change in temperature from 0.5°C to 4°C under most climate change scenarios drawn from the special report on emission scenarios (SRES) and nine climate models (Luo, Bellotti, Williams, & Bryan, 2005).

Since climate changes are different in different agroclimatic zones, therefore, impact will be different depending upon the cropping systems, management interventions, and soil type. Jalota and Vashisht (2016) projected that temperature (average of three general circulation models

(GCMs)—HadCM3, CSIRO, and CCCMA—and two SERES scenarios, A2 and B2) in 2020, 2050, and 2080 years would increase by 1.4°C, 2.7°C, and 3.9°C in hot humid (maize-wheat cropping system); 1.6°C, 2.4°C, and 3.5°C in hot subtropical (rice-wheat cropping system); and 1.4°C, 2.6°C, and 3.7°C in hot arid (cotton-wheat cropping system) zones of Indian Punjab. With the projected increased temperatures in 2020, 2050, and 2080, the crop duration would be shortened by 18, 30, and 42 days in maize-wheat; 18, 32, and 44 days in rice-wheat; and 18, 34, and 48 days in cotton-wheat cropping systems in respective years. The reduction in yield of maize-wheat, rice-wheat, and cotton-wheat cropping systems would be almost 0%, 3.8%, and 10.2% in 2020; 4.2%, 8.0%, and 21.2% in 2050; and 13.4%, 17.1%, and 33.1% in 2080, respectively. The projected trends of climate data with regional climate model (PRECIS) and its impact on rice-wheat system are similar to that of GCM but of different magnitude. With an increase in temperature by 2.4°C in MC and by 5.1°C in EC compared with the present time slice in hot subtropical climate of central Punjab, it is projected that the crop duration of rice would be shortened by 19 and 26 days; of wheat by 9 and 23 days in MC and EC, respectively and accordingly, there would be a reduction in yield (Fig. 3.4).

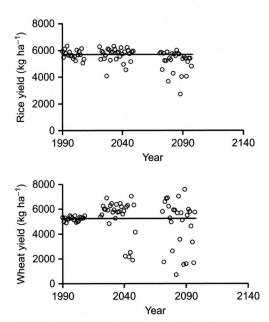

Fig. 3.4 Yields of rice and wheat crops in present time slice (1989–2010), mid century (2021–50), and end century (2071–98).

3.12 EFFECT OF CLIMATE CHANGE ON ROOT-ZONE WATER BALANCE

Climate and water resources are closely interlinked. As a first response to climate trends, it is expected that projected changes in CO_2, temperature, and P_{cp} at local/regional/global scale will alter root-zone water-balance components and groundwater recharge/draft from aquifers, causing shifts in groundwater levels in aquifers. High natural variability in climate and catchment properties (slope, soil type, depth of soil, land use, etc.) can cause variations in amount and distribution of the soil water in root zone of the cropped area. The water balance in root zone of the cropped area is represented by Eq. (3.11):

$$I + P_{cp} = E_c + E_b + T + R + \Delta SWS + D \qquad (3.11)$$

where I is irrigation, P_{cp} is precipitation, E_c is soil water evaporation during cropped period, E_b is soil water evaporation from noncropped or fallow period (s), T is transpiration from the canopy, D is the drainage beyond root zone, R is surface runoff water, and ΔSWS is change in soil water storage in the root zone.

Changes in total and distribution of seasonal P_{cp} and temperature are important for agriculture as these strongly influence the availability of water to crops. Temperature and P_{cp} affect E and T components directly and ΔSWS and D indirectly, as a result of change in ET. Extreme P_{cp} events cause more runoff and increase soil erosion, the major land degradation process. Elevated temperature due to climate change leads to intensification of the water loses in hydrologic cycle by increasing soil water evaporation and accelerating transpiration in the plants. Higher ET causes soil moisture stress leading to plant growth reduction, which gets reflected in plant height, leaf area, dry weight, and other growth functions. Soil moisture stress during flowering, pollination, and grain filling retards plant development and subsequently decreases yield in most of the crops. Due to increased ET and water stress, the demand of groundwater for irrigation is expected to rise in a warmer climate, bringing increased competition between agriculture (already the largest consumer of water resources in semiarid regions), urban, and industrial users. Intensified evaporation also increases the hazard of salt accumulation in the surface soil. With an increase in temperature though ET rate is increased, total ET is decreased during a crop season. The decrease is not in both the components, that is, E and T, but is only in the latter (T).

Though by increased temperature transpiration rate is stimulated owing to more vapor pressure deficit (Eq. 3.4), the total transpiration is decreased due to shortening of crop duration (Jalota, Kaur, Kaur, et al., 2013). The T component of water balance is also influenced by atmospheric CO_2 in addition to temperature and P_{cp}. Transpiration is decreased with increased CO_2 due to partial closure of stomatal aperture.

In cropped soils, water-balance components are influenced jointly by changes in temperature, CO_2, and P_{cp} and their variability. The assessment of such effects for the projected climate change scenario is difficult by experimentation; alternately, simulations are preferred. In simulation studies, Jalota, Kaur, Kaur, et al. (2013) projected that with increased temperature, CO_2, and P_{cp} during MC and EC in semiarid region of India, crop duration and water-balance components would be different as compared with present time slice (Table 3.10). In China, a decrease in ET of wheat would range from −2.0% to 14.0% and of maize from −3.0% to 21.8% under future climate change projections for three periods, that is, 2030–39, 2060–69, and 2090–99 (Guo, Lin, Mo, & Yang, 2010). In the future, though ET would decrease, the E component would increase by 29% and 23% in rice and 64% and 56% in wheat during MC and EC, respectively (Table 3.10). The effect of climate change on water-balance components would also vary with soil texture. The change in soil water evaporation would be comparatively more in fine-textured soils. Water-use efficiency (grain yield/ET) in rice (averaged across soil type) would decrease by 5% in both MC and EC. However, in wheat, the WUE would increase by 4% in MC and 18% in EC. Contrary in areas where temperature is less than the optimum range, increased temperature by climate change not only increases crop yields but also has positive effect on WUE, mainly ascribed to the evapotranspiration intensification (Guo et al., 2010). They projected that WUE of wheat would improve 28.1% in the 2030s, 55.8% in the 2060s, and 78.1% in the 2090s.

It is not only the WUE but also the water productivity (yield/irrigation water) that would increase in the future owing to reduction in irrigation water requirement through shortening of crop duration with increased temperature, less transpiration with increased CO_2, and more water supply from increased P_{cp}. In the future, the extent of water-balance components would also vary spatially and temporally (Jalota et al., 2014). For example, in the central Punjab of India during rice-wheat cropping period with higher P_{cp} at Jalandhar in PTS and at Patiala in MC and EC, I, T, and ET at Amritsar and E at Patiala remained higher than other locations, irrespective of the time slice (Table 3.11). More drainage at Jalandhar and higher WUE efficiency at Ludhiana at PTS would shift to Ludhiana and Jalandhar,

Table 3.10 Percent Change in Yield, Water-Balance Components, Water-Use Efficiency, and Water Productivity of Rice and Wheat Crops as Influenced by Soil Texture in Mid and End of the 21st Century

Soil	Time Slice	Yield	I	P_{cp}	ET	E	T	D	ΔSWS	WUE	WP
						Rice					
sl	MC	4.9	48.3	−32.3	8.6	−25.8	23.3	34.0	−16.7	−3.5	−83.9
	EC	14.0	67.8	−55.8	18.6	−20.1	34.9	13.6	93.2	−5.3	−163.7
sil	MC	22.3	43.9	−32.3	9.1	−32.2	24.8	25.2	−15.6	14.3	−38.9
	EC	32.8	65.3	−55.8	18.7	−25.0	37.7	−0.9	94.8	15.6	−93.7
						Wheat					
sl	MC	−2.4	21.5	−134.0	1.5	−80.7	22.3	−15.0	−169.2	−4.0	−30.4
	EC	9.2	56.3	−102.9	26.1	−71.1	50.8	−28.8	165.4	−23.1	−108.4
sil	MC	2.9	24.5	−134.0	6.0	−47.8	26.9	−433.3	−275.0	−3.1	−35.4
	EC	16.2	60.5	−102.9	25.4	−41.8	60.4	−700.0	343.8	−12.2	−138.6

I, P_{cp}, ET, E, T, D, ΔSWS, WUE, and WP represent irrigation, precipitation, evapotranspiration, evaporation from soil, transpiration, drainage, change in soil water storage in 0–1.8 m soil depth, water-use efficiency, and water productivity, respectively. sl and sil denote sandy loam and silt loam soils. MC and EC represent the mid (2021–50) and end (2071–98) century time slices.
Negative values represent the increase.

Table 3.11 Water Balance and Water-Use Efficiency in Rice-Wheat Cropping System in Different Time Slices of the 21st Century at Different Locations in Central Punjab

Location	I	P_{cp}	ET	E	T	D	ΔS	WUE ($kg\,m^{-3}$)
				mm				
Present time slice (1989–2010)								
Amritsar	1709	716	1250	424	826	1246	−71	0.98
Jalandhar	1424	977	1155	395	759	1285	−39	1.02
Ludhiana	1527	757	1076	382	695	1252	−44	1.07
Patiala	1519	790	1137	436	701	1183	−11	1.04
Mid century (2021–50)								
Amritsar	1054	879	1131	564	567	841	−39	1.03
Jalandhar	849	969	1015	492	523	785	19	1.10
Ludhiana	874	1053	1052	523	528	881	−5	1.00
Patiala	774	1105	1070	581	488	856	−46	1.08
End century (2071–98)								
Amritsar	1046	907	1021	601	419	971	−38	0.91
Jalandhar	820	1172	937	543	395	1067	−13	1.11
Ludhiana	807	1210	962	551	411	1067	−12	1.08
Patiala	702	1321	1001	612	389	1051	−29	0.90

I, P_{cp}, ET, E, T, D, ΔS, and WUE represent irrigation, precipitation, evapotranspiration, evaporation from soil, transpiration, drainage, change in soil water storage in 0–1.8 m soil depth, and water-use efficiency.

respectively, in MC and EC. In addition to location, the magnitude of root-zone water-balance components is likely to vary with season and land use owing to shrinkage of agricultural lands and swallowing of urbanization in the future. The higher contribution of I, E, and D (on volume basis) from agricultural lands and of R from water bodies than other land uses during PTS in both seasons *Kharif* (local name of a cropping season, from June to October) and *Rabi* (local name of a cropping season, from November to May) would change in the future.

EXERCISES

1. Calculate the increase in temperature that will nullify the benefit effect of doubling of CO_2 in rice, wheat, cotton, and maize crops using the coefficients from Table 3.7 and the model as Eq. (3.9), given in this chapter. Assume present CO_2, mean maximum temperature, and mean minimum temperature are 380 ppm, 24.0°C, and 10.2°C, respectively.

   Answer (1.1°C, 1.7°C, 0.6°C, and 0.5°C in rice, wheat, cotton, and maize crops, respectively).

2. (a) How do elevated CO_2, temperature, precipitation, and their interactions affect the crop productivity?

(b) How these effects are modified by photosynthetic pathway of the plant, soil water regime, and soil nitrogen level?

3. Describe the plant and soil processes by which climate parameters affect the root-zone water-balance components.

4. By which means does rising CO_2 concentration make it easier to the plants for coping water and nutrient stresses?

5. Under which conditions does photosynthesis in C4 plants respond to elevated CO_2 and why? Will additional nitrogen fertilizers applied to maize crop increase yield, NUE, and quality in the coming years?

REFERENCES

Abe, I. (1969). Agro-meterological studies regarding the effect of early winds ("Yamase wind") on growth of rice plants in Amomori prefecture. *Bulletin of Aomori Agricultural Experiment Station, 14*, 40–138 [In Japanese].

Abrol, Y. P., Bagga, N., Chakravarty, N. V. K., & Wattal, P. N. (1991). In Y. P. Abrol, A. Gnanam, D. R. Govindjee, A. H. Ort Teramura, & P. N. Wattal (Eds.), *Impact of rise in temperature on the productivity of wheat in India. Proceedings of the Indo-US workshop on impact of global climate change on photosynthesis and plant productivity* (pp. 787–798). New Delhi: Oxford and IBH Publishing.

Acock, B., & Allen, L. H., Jr. (1985). Crop responses to elevated carbon dioxide concentrations. In B. R. Strain & J. D. Cure (Eds.), *Direct effects of increasing carbon dioxide on vegetation* (pp. 317–346). Washington, DC: Office of Energy Research, U.S. Dept. of Energy. DOE/ER–0238.

Aggarwal, P. K., & Kalra, N. (1994). *Simulating the effect of climate factors, genotype and management on productivity of wheat in India.* New Delhi, India: Indian Agricultural Research Institute Publication [pp. 156].

Aggarwal, P. K., & Sinha, S. K. (1993). Effect of probable increase in carbon dioxide and temperature on productivity of wheat in India. *Journal of Agricultural Meteorology, 48*, 811–814.

Ainsworth, E. (2008). A Rice production in a changing climate: A meta-analysis of responses to elevated carbon dioxide and elevated ozone concentration. *Global Change Biology, 14*, 1642–1650.

Ainsworth, E. A., Davey, P. A., Bernacchi, C. J., Dermody, O. C., Heaton, E. A., Moore, D. J., et al. (2002). A meta analysis of elevated CO_2 effects on soybean (*Glycine max*) physiology, growth and yield. *Global Change Biology, 8*, 695–709.

Ainsworth, E. A., & Long, S. P. (2005). What have we learned from 15 years of free-air CO_2 enrichment (FACE)? A meta-analytic review of the responses of photosynthesis, canopy properties and plant production to rising CO_2. *New Phytologist, 165*, 351–372.

Ainsworth, E. A., & Rogers, A. (2007). The response of photosynthesis and stomatal conductance to rising (CO_2): Mechanisms and environmental interactions. *Plant, Cell and Environment, 30*, 258–270.

Allen, L. H., Jr. (1994). Direct impact on crops and indirect effects mediated through anticipated climatic changes. In K. J. Boote, T. R. Sinclair, & J. M. Bennett (Eds.), *Physiology and determination of crop yield, Carbon dioxide increase* (pp. 425–459). Madison, WI: ASA, CSSA, SSSA.

Allen, L. H., Jr., Pan, D., Boote, K. J., Pickering, N. B., & Jones, J. W. (2003). Carbon dioxide and temperature effects on evapotranspiration and water-use efficiency of soybean. *Journal of Agronomy, 95*, 1071–1081.

Andre, M., & Du Cloux, H. (1993). Interaction of CO_2 enrichment and water limitations on photosynthesis and water use efficiency in wheat. *Plant Physiology and Biochemistry*, *31*, 103–112.

Asseng, S., Jamieson, P. D., Kimball, B., Pinter, P., Sayre, K., Bowden, J. W., et al. (2004). Simulated wheat growth affected by rising temperature, increased water deficit and elevated CO_2. *Field Crops Research*, *85*, 85–102.

Baker, J. T., Allen, L. H., Jr., & Boote, K. J. (1990). Growth and yield response to rice to CO_2 concentration. *Journal of Agricultural Science*, *115*, 113–120.

Baker, J. T., Allen, L. H., Jr., & Boote, K. J. (1992). Temperature effects on rice at elevated CO_2 concentrations. *Journal of Experimental Botany*, *43*, 959–964.

Baker, J. T., Allen, L. H., Jr., & Boote, K. J. (1995). Potential climate change effects on rice: Carbon dioxide and temperature. In C. Rosenzweig, L. A. Harper, S. E. Hollinger, J. W. Jones, & L. H. Allen Jr., (Eds.), *Climate change and agriculture: Analysis of potential international impacts* (pp. 31–47). Madison, WI: American Society of Agronomy. ASA Special publication No. 59.

Baker, J. T., Allen, L. H., Jr., Boote, K. J., Jones, P., & Jones, J. W. (1989). Response of soybean to air temperature and carbon dioxide concentration. *Crop Science*, *29*, 98–105.

Bannayan, M., & Sanjani, S. (2011). Weather conditions associated with irrigated crops in an arid and semi-arid environment. *Agricultural and Forest Meteorology*, *151*, 1589–1598.

Batts, G. R., Morrison, J. I. T., Eliis, R. H., Hadley, P., & Wheeler, T. R. (1997). Effects of CO_2 and temperature on growth and yield of crop of winter wheat over four seasons. *European Journal of Agronomy*, *7*, 43–52.

Bender, J., Herstein, U., & Black, C. R. (1999). Growth and yield responses of spring wheat to increasing carbon dioxide, ozone and physiological stresses: A statistical analysis of "ESPACE-wheat" results. *European Journal of Agronomy*, *10*, 185–195.

Bernacchi, C. J., Kimball, B. A., Quarles, D. R., Long, S. P., & Ort, D. R. (2007). Decreases in stomatal conductance of soybean under open-air elevation of CO_2 are closely coupled with decreases in ecosystem evapotranspiration. *Plant Physiology*, *143*, 134–144.

Bloom, A. J., Burger, M., Rubio Asensio, J. S., & Cousins, A. B. (2010). Carbon dioxide inhibits nitrate assimilation in wheat and Arabidopsis. *Science*, *328*, 899–903.

Boote, K. J., Jones, J. W., & Hoogenboom, G. (1998). Simulation of crop growth: CRPGRO Model. In R. M. Pearr & R. B. Curry (Eds.), *Agricultural systems modeling and simulation* (pp. 651–692). New York: Marcel Dekker.

Boote, K. J., Pickering, N. B., & Allen, L. H., Jr. (1997). Plant modeling: Advances and gaps in our capability to project future crop growth and yield in response to global climate change. In L. H. Allen, Jr., M. B. Kirkham, C. Whitman, & D. M. Olzyk (Eds.), *Advances in carbon dioxide research* (pp. 179–228). Madison, WI: American Society of Agronomy. ASA Special publication No. 61.

Brown, R. A., & Rosenberg, N. J. (1999). Climate change impacts on the potential productivity of corn and winter wheat in their primary United States growing regions. *Climate Change*, *41*, 73–107.

Bunce, J. A. (1995). Long term growth of alfalfa and orchard grass plots at elevated carbon dioxide. *Journal of Biogeography*, *22*, 341–348.

Bunce, J. A. (2000). Response of stomatal conductance to light, humidity and temperature in winter wheat and barley grown at three concentrations of carbon dioxide in the field. *Global Change Biology*, *6*, 371–382.

Burkart, S., Manderscheid, R., Wittich, K. P., Lopmeier, F. J., & Weigel, H. J. (2011). Elevated CO_2 effects on canopy and soil water flux parameters measured using a large chamber in crops grown with free-air CO_2 enrichment. *Plant Biology*, *13*, 258–269.

Buttar, G. S., Jalota, S. K., Sood, A., & Bhushan, B. (2012). Yield and water productivity of Bt cotton (*Gossupium hirsutum*) as influenced by temperature under semi-arid conditions of north-western India: field and simulation study. *Indian Journal of Agricultural Sciences*, *82*, 44–49.

Chahal, G. B. S., Sood, A., Jalota, S. K., Choudhury, B. U., & Sharma, P. K. (2007). Yield, evapotranspiration and water productivity of rice (*Oryza sativa* L.)—wheat (*Triticum aestivum* L.) system in Punjab (India) as influenced by transplanting date of rice and weather parameters. *Agricultural Water Management, 88*, 14–22.

Chaterjee, A. (1998). *Simulating the effect of increase in temperature and CO_2 on growth and yield of maize and sorghum* [MSc thesis]. New Delhi: Indian Agricultural Research Institute.

Cheng, W., & Kuzyakov, Y. (2005). Root effects on soil organic matter decomposition. *American Society of Agronomy, 13*, 119–143.

Chowdhury, S. I. C., & Wardlaw, I. F. (1978). The effect of temperature on kernel development in cereals. *Australian Journal of Agricultural Research, 29*, 205–233.

Coles, G. D., Hartunian-Sowa, S. M., Jamieson, P. D., Hay, A. J., Atwell, W. A., & Fulcher, R. G. (1997). Environmentally induced variation in starch and non-starched poly-saccharides content in wheat. *Journal of Cereal Science, 26*, 47–54.

Conley, M. M., Kimball, B. A., Brooks, T. J., Pinter, P. J., Jr., Hunsaker, D. J., Wall, G. W., et al. (2001). CO_2 enrichment increases water use efficiency in sorghum. *New Phytologist, 151*, 407–412.

Cooper, N. T. W., Siebenmorgen, T. J., & Counce, P. A. (2008). Effects of night time temperature during kernel development on rice physicochemical properties. *Cereal Chemistry, 85*, 276–282.

Cotrufo, M. F., Ineson, P., & Scott, A. (1998). Elevated CO_2 reduces the nitrogen concentration of plant tissues. *Global Change Biology, 4*, 43–54.

Cousins, A. B., Adam, N. R., Wall, G. W., Kimball, B. A., Printer, P. J., Ottman, M. J., et al. (2003). Development of C4 photosynthesis in sorghum leaves grown under free-air CO_2 enrichment (FACE). *Journal of Experimental Botany, 54*, 1969–1975.

Cuculeanu, B., Marica, A., & Simota, C. (1999). Climate change impact on agriculture crops and adaptation options in Romania. *Climate Research, 12*, 153–160.

Das, L., Lohar, D., Sadhukhan, L., Khan, S. A., Saha, A., & Sarkar, S. (2007). Evaluation of the performance of ORYZA 2000 and assessing the impact of climate change on rice production in Gangetic West Bengal. *Journal of Agrometeorology, 9*, 1–10.

De Costa, W. A. J. M., Weerakoon, W. M. W., Herath, H. M. L. K., Amaratunga, S. P., & Abeywardena, R. M. I. (2006). Physiology of yield determination of rice under elevated carbon dioxide at high temperatures in a subhumid tropical climate. *Field Crops Research, 96*, 336–347.

De Graaff, M. A., Van Groenigen, K. J., Six, J., Hungate, B., & Van Kessel, C. (2006). Interactions between plant growth and soil nutrient cycling under elevated CO_2: A meta-analysis. *Global Change Biology, 12*, 2077–2091.

De Siqueira, O. J. F., Farias, J. R. F., & Sans, L. M. A. (1994). Potential effect of global climate change for Brazilian agriculture: Applied simulation studies for wheat, maize and soybean. In C. Rosenzweig & A. Iglesias (Eds.), *Implications of climate change for international agriculture: Crop modeling study*. Washington, DC: US Environmental protection Agency. EPA2 230-B-94-003.

Diaz, S., Grime, J. P., Harris, J., & McPherson, E. (1993). Evidence of a feedback mechanism limiting plant response to elevated carbon dioxide. *Nature, 364*, 616–617.

Erbs, M., Manderscheid, R., Jensen, G., Seddig, S., Pacholski, A., & Weigel, H. J. (2010). Effects of free air CO_2 enrichment and nitrogen supply on grain quality parameters of wheat and barley grown in a crop rotation. *Agriculture, Ecosystems and Environment, 136*, 59–68.

Evans, J. (2009). 21st century Climate Change in the Middle East. *Climate Change, 92*, 417–432.

Ferris, R., Wheeler, T. R., Hadley, P., & Ellis, R. H. (1998). Recovery of photosynthesis after environmental stress in Soybean grown under elevated CO_2. *Crop Science, 38*, 948–955.

Field, C. B., Jackson, R. B., & Mooney, H. A. (1995). Stomatal response to increased CO_2: Implications from plant to global scale. *Plant, Cell and Environment, 18*, 1214–1225.

Finzi, A. C., Sinsabaugh, R. L., Long, T. M., & Osgood, M. P. (2006). Microbial community responses to atmospheric carbon dioxide enrichment in a warm-temperate forest. *Ecosystems, 9*, 215–226.

Furbank, R. T., & Hatch, M. D. (1987). Mechanism of C4 photosynthesis: The size and composition of inorganic pool in bundle sheet cells. *Plant Physiology, 85*, 958–964.

Gaastra, P. (1962). Photosynthesis of leaves and field crops (sic). *Netherlands Journal of Agricultural Science, 10*, 311–324.

Ghaffari, A., Cook, H. F., & Lee, C. (2002). Climate change and winter wheat management: A modelling scenario for south-eastern England. *Climatic Change, 55*, 509–533.

Gifford, R. M. (2004). The CO_2 fertilising effect-does it occur in real world? The international free air enrichment (FACE) workshop: Short and long term effects of elevated atmospheric CO_2 on managed ecosystem, Ascona Switzerland. *New Phytologist, 163*, 221–225.

Gill, S. S., & Tuteja, N. (2010). Reactive oxygen species and antioxidant machinery in abiotic stress tolerance in crop plants. *Plant Physiology and Biochemistry, 48*, 909–930.

Grayston, S. J., Campbell, C. D., Lutze, J. L., & Gifford, R. M. (1998). Impact of elevated CO_2 on the metabolic diversity of microbial communities in N limited grass swards. *Plant and Soil, 203*, 289–300.

Guo, R. P., Lin, Z., Mo, X., & Yang, C. (2010). Responses of crop yield and water use efficiency to climate change in the North China Plain. *Agricultural Water Management, 97*, 1185–1194.

Hakala, K. (1998). Growth and yield potential of spring wheat in a simulated changed climate with increased CO_2 and higher temperature. *European Journal of Agronomy, 9*, 41–52.

Hatfield, J. L., Boote, K. J., Kimball, B. A., Ziska, L. H., Izaurralde, R. C., Ort, D., et al. (2011). Climate change impact on agriculture: Implications for crop production. *Agronomy Journal, 103*, 351–370.

Hocking, P. J., & Meyer, C. P. (1991). Effects of CO_2 enrichment and nitrogen stress on growth and partitioning of dry matter and nitrogen in wheat and maize. *Australian Journal of Plant Physiology, 18*, 339–356.

Horrie, T., Baker, J. T., Nakagawa, H., Matsui, T., & Kim, H. Y. (2000). Crop ecosystem responses to climatic change: Rice. In K. R. Reddy & H. F. Hodges (Eds.), *Climate change and global crop productivity* (pp. 81–106). New York: CAB International.

Huang, J. J., & Lur, H. S. (2000). Influences of temperature during grain filling stages on grain quality in rice (*Oryza sativa* L.). 1. Effects of temperature on yield components, milling quality, and grain physicochemical properties. *Journal of the Agriculture Association of China, 1*, 370–389.

Hundal, S. S., & Kaur, P. (1996). Climatic change and its impacts on crop productivity in Punjab, India. In Y. P. Abrol, S. Gadgil, & G. B. Pant (Eds.), *Climatic variability and agriculture* (pp. 377–393). New Delhi: Narosa Publishing House.

Hundal, S. S., & Kaur, P. (2007). Climatic variability and its impact on cereal productivity in Indian Punjab. *Current Science, 92*, 506–512.

Hurkman, W. J., Vensel, W. H., Tanaka, C. K., Whitehand, L., & Altenbach, S. B. (2009). Effects of high temperature on albumin and globulin accumulation in endosperm proteome of the developing wheat grain. *Journal of Cereal Science, 49*, 12–23.

Hussain, S. S., & Mudassar, M. (2007). Prospects for wheat production under changing climate in mountain area of Pakistan—An econometric analysis. *Agricultural Systems, 94*, 494–501.

Iglesias, A., & Minguez, M. I. (1997). Modelling crop climate interactions in Spain vulnerability and adaptation of different agricultural systems to climate change. *Mitigation and Adaptation Strategies for Global Change, 1*, 273–288.

IPCC. (2001). *Climate change 2001: Impacts, adaptation and vulnerability.* Summary for policy makers. Assessment Report of the Intergovernmental Panel on Climate Change Cambridge: Cambridge University Press.

IPCC. (2007). Impacts, adaptation and vulnerability. Contribution of Working group II to the Intergovernmental Panel on Climate Change. In M. L. Parry, O. F. Canziani, J. P. Palutikof, P. J. van der Linden, & C. E. Hansen (Eds.), *Fourth assessment report.* New York: Cambridge University Press.

IPCC. (2011). *Extreme weather and climate change.* http://climate.uu-uno.org/articles/view/171595.

IPCC. (2014). Summary for policymakers, climate change 2014: Impacts, adaptation, and vulnerability. Part A: Global and sectoral aspects. Working group II contribution to the Intergovernmental panel on climate change. In C. B. Field, V. R. Barros, D. J. Dokken, K. J. Mach, M. D. Mastrandrea, T. E. Bilir, M. Chatterjee, K. L. Ebi, Y. O. Estrada, R. C. Genova, B. Girma, E. S. Kissel, A. N. Levy, S. MacCracken, P. R. Mastrandrea, & L. L. White (Eds.), *Fifth Assessment Report* (pp. 1–32). Cambridge, NewYork: Cambridge University Press.

Izaurralde, R. C., Thomson, A. M., Morgan, J. A., Fay, P. A., Polley, H. W., & Hatfield, J. L. (2011). Climate impacts on agriculture: Implications for Forage and Rangeland production. *Journal of Agronomy, 103*, 371–380.

Jablonski, L. M., Wang, X., & Curtis, P. S. (2002). Plant reproduction under elevated CO_2 conditions: A meta-analysis of reports on 79 crop and wild species. *New Phytologist, 156*, 9–26.

Jalota, S. K., Kaur, H., Kaur, S., & Vashisht, B. B. (2013). Impact of climate change scenario on yield, water and nitrogen-balance and -use efficiency of rice-wheat cropping system. *Agricultural Water Management, 116*, 29–38.

Jalota, S. K., Kaur, H., Ray, S. S., Tripathy, R., Vashisht, B. B., & Bal, S. K. (2013). Past and general circulation model driven future trends of climate change in central Indian Punjab: Ensuing yield of rice-wheat cropping system. *Current Science, 104*, 105–110.

Jalota, S. K., Ray, S. S., & Panigrahy, S. (2009). In P. Shushma, S. R. Shibendu, & P. Jai Singh (Eds.), *Effect of elevated CO_2 and temperature on productivity of three main cropping systems in Punjab state of India: A simulation analysis. Impact of climate change on agriculture (pp. 135–142). Workshop proceedings, ISPRS Arcives XXXVIII–8/W20.*

Jalota, S. K., Sukhwinder, S., Chahal, G. B. S., Ray, S. S., Panigrahy, S., Bhupinder, S., et al. (2010). Soil texture, climate and management effects on plant growth, grain yield and water use by rained maize (*Zea mays* L.)—wheat (*Triticum aestivum* L.) cropping system: Field and simulation study. *Agricultural Water Management, 97*, 83–90.

Jalota, S. K., & Vashisht, B. B. (2016). Adapting cropping systems to future climate change scenario in three agro-climatic zones of Punjab, India. *Journal of Agrometeorology, 18*, 48–56.

Jalota, S. K., Vashisht, B. B., Kaur, H., Kaur, S., & Kaur, P. (2014). Location specific climate change scenario and its impact on rice-wheat cropping system. *Agricultural Systems, 131*, 77–86.

Jones, P., Allen, L. H., Jr., & Jones, J. W. (1985). Response of Soybean canopy photosynthesis and transpiration to whole day temperature changes in different CO_2 environments. *Agronomy Journal, 77*, 242–249.

Jones, P., Allen, L. H., Jr., Jones, J. W., & Valle, R. (1985). Photosynthesis and transpiration responses of soybean canopies to short-and long-term CO_2 treatments. *Agronomy Journal, 77*, 119–126.

Jones, P. G., & Thornton, P. K. (2003). The potential impacts of climate change in tropical agriculture: The case of maize in Africa and Latin America in 2055. *Global Environmental Change, 13*, 51–59.

Jungk, A. O. (2002). Dynamics of nutrient movement at soil-root interface. In Y. Waisel, A. Eshel, & U. Kafkafi (Eds.), *Plant roots: The hidden half* (3rd ed., pp. 587–616). New York: Marcel Dekker.

Kang, S., Zhang, F., Hu, X., & Zhang, J. (2002). Benefits of CO_2 enrichment on crop plants are modified by soil water stress. *Plant and Soil, 238*, 69–77.

Kaur, P., & Hundal, S. S. (2006). Effect of possible futuristic climate change scenarios on productivity of some kharif and rabi crops in the central agroclimatic zone of Punjab. *Journal of Agricultural Physics, 6*, 21–27.

Kaur, P., & Hundal, S. S. (2010). Global climate change vis-à-vis crop productivity. In M. K. Jha (Ed.), *Natural and anthropogenic disasters—Vulnerability, preparedness and mitigation* (pp. 413–431). New Delhi/The Netherlands: Capital Publishing Company/ Springer.

Kaur, H., Jalota, S. K., Kanwar, R., & Vashisht, B. B. (2012). Climate change impacts on yield, evapotranspiration and nitrogen uptake in irrigated maize (*Zea mays*)—wheat (*Triticum aestivum*) cropping system: A simulation analysis. *Indian Journal of Agricultural Sciences, 82*, 213–219.

Kellither, F. M., Leuning, R., Raupach, M. R., & Schelze, E. D. (1995). Maximum conductance for evaporation from global vegetation type. *Agricultural and Forest Meteorology, 73*, 1–16.

Kim, H. Y., Horie, T., Nakagawa, H., & Wada, K. (1996). Effects of elevated CO_2 concentration and high temperature on growth and yield of rice II.The effect on yield and its components of Akihikari rice. *Japanese Journal of Crop Science, 65*, 644–651.

Kim, H. Y., Lieffering, M., Kobayashi, K., Okada, M., Mitchel, M. W., & Gumpertz, M. (2003). Effects of free air CO_2 enrichment and nitrogen supply on the yield of temperate paddy rice crops. *Field Crops Research, 82*, 261–270.

Kimball, B. A. (1983). Carbon dioxide and agricultural yield: An assemblage and analysis of 430 prior observations. *Agronomy Journal, 75*, 779–788.

Kimball, B. A., & Idso, S. B. (1983). Increasing atmospheric CO_2: Effects on water use, crop yield and climate. *Agricultural Water Management, 7*, 55–72.

Kimball, B. A., LaMorte, R. L., Pinter, P. J., Jr., Wall, G. W., Hunsaker, D. J., Adamsen, F. J., et al. (1999). Free-air CO_2 enrichment (FACE) and soil nitrogen effects on energy balance and evapotranspiration of wheat. *Water Resources Research, 35*, 1179–1190.

Kimball, B. A., & Mauney, J. R. (1993). Response of cotton to varying CO_2, irrigation and nitrogen: Yield and growth. *Agronomy Journal, 85*, 706–712.

Kimball, B. A., Morris, C. F., Pinter, P. J., Jr., Wall, G. W., Hunsaker, D. J., Adamsen, F. J., et al. (2001). Elevated CO_2, drought and soil nitrogen effects on wheat quality. *New Phytologist, 150*, 295–303.

Kimball, B. A., Pinter, P. J., Garcia, R. L., LaMorte, R. L., Wall, G. W., Hunsaker, D. J., et al. (1995). Productivity and water use of wheat under free-air CO_2 enrichment. *Global Change Biology, 1*, 429–442.

Kirkham, M. B. (2011). *Elevated carbon dioxide: Impacts on soil and plant water relations*. Boca Raton, FL: CRC Press, Taylor and Francis Group.

Knox, J. W., Hess, T. M., Daccache, A., & PreezOrtola, M. (2011). *What are the projected impacts of climate change on food crop productivity in Africa and South Asia? DFID systematic Review*. Final Report, Cranfield University, (pp. 77).

Kohler, J., Hernandez, J. A., Caravaca, F., & Roldan, A. (2009). Induction of antioxidant enzymes is involved in the greater effectiveness of a PGPR versus AM fungi with respect to increasing the tolerance of lettuce to severe salt stress. *Environmental and Experimental Botany, 65*, 245–252.

Kooman, P. L. (1995). *Yielding ability of potato crops as influenced by temperature and daylength* [PhD dissertation]. Wageningen, The Netherlands: Wageningen Agricultural University.

Koranda, M., Schnecker, J., Kaiser, C., Fuchslueger, L., Kitzler, B., Stange, C. L., et al. (2011). Microbial processes and community composition in the rhizosphere of European beech the influence of plant C exudates. *Soil Biology and Biochemistry, 43*, 551–558.

Kucharik, C. J., & Serbin, S. P. (2008). Impacts of recent climate change on Wisconsin corn and soybean yield trends. *Environmental Research Letters, 3*, 1–10.

Kuzyakov, Y., & Xu, X. L. (2013). Tansley review: Competition between roots and microorganisms for nitrogen: Mechanisms and ecological relevance. *New Phytologist, 198*, 656–669.

Lal, M., Singh, K. K., Rathore, L. S., Srinivasan, G., & Saseendran, S. A. (1998). Vulnerability of rice and wheat yields in the NW India to future changes in climate. *Agricultural and Forest Meteorology, 89*, 101–114.

Lal, M., Singh, K. K., Srinivasan, G., Rathore, L. S., Naidu, D., & Tripathi, C. N. (1999). Growth and yield responses of soybean in Madhya Pradesh, India to climate variability and change. *Agricultural and Forest Meteorology, 93*, 53–70.

Lesk, C., Rowhan, P., & Ramankutty, N. (2016). Influence of extreme weather disasters on global crop production. *Nature, 529*, 84–87. 16467.

Li, W., Han, X., Zhang, Y., & Li, Z. (2007). Effects of elevated CO_2 concentration, irrigation and nitrogenous fertilizer application on the growth and yield of spring wheat in semi-arid areas. *Agricultural Water Management, 87*, 106–114.

Li, F., Kang, S., & Zhang, J. (2004). Interactive effect of elevated CO_2, nitrogen and drought on leaf area, stomatal conductance and evapotranspiration of wheat. *Agricultural Water Management, 67*, 221–233.

Li, M., Li, Z., Wang, D., Zhang, X., Li, Z., & Li, Y. (2005). Impact of natural disasters change on grain, yield in China in the past 50 years. *Journal of Natural Disasters, 14*, 55–60.

Lin, S. K., Chang, M. C., Tsai, Y. G., & Lur, H. S. (2005). Proteomic analysis of the expression of proteins related to rice quality during caryopsis development and the effect of high temperature on expression. *Proteomics, 5*, 2140–2156.

Lin, C. J., Li, C. Y., Lin, S. K., Yang, F. H., Huang, J. J., Liu, Y. H., et al. (2010). Influence of high temperature during grain filling on the accumulation of storage proteins and grain quality in rice (*Oryza sativa* L.). *Journal of Agricultural and Food Chemistry, 58*, 10545–10552.

Lobell, D. B., & Field, C. B. (2007). Global scale climate-crop yield relationships and impact of recent warming. *Environmental Research Letters, 2*, 1–7.

Lobell, D. B., Schlenker, W., & Costa-Roberts, C. (2011). Climate trends and global crop production since 1980. *Science, 333*, 616–620.

Loladze, I. (2002). Rising atmospheric CO_2 and human nutrition: Toward globally imbalanced plant stoichiometry? *Trends in Ecology & Evolution, 17*, 457–461.

Long, S. P. (1999). Environmental responses. In R. F. Sage & R. K. Monson (Eds.), *C4 Plant biology* (pp. 215–249). San Diego, CA: Academic Press.

Long, S. P., Ainsworth, E. A., Leakey, A. D. B., Nosberger, J., & Ort, D. R. (2006). Food for thought: Lower-than-expected crop yield stimulation with rising CO_2 concentrations. *Science, 312*, 1918–1921.

Long, S. P., Ainsworth, E. A., Rogers, A., & Ort, D. R. (2004). Rising atmospheric carbon dioxide: Plants face the future. *Annual Review of Plant Biology, 55*, 591–628.

Ludwig, F., & Asseng, S. (2006). Climate change impacts on wheat production in a Mediterranean environment in Western Australia. *Agricultural Systems, 90*, 159–179.

Luo, Q. Y., Bellotti, W., Williams, M., & Bryan, B. (2005). Potential impact of climate change on wheat yield in South Australia. *Agricultural and Forest Meteorology, 132*, 273–285.

Mall, R. K., Singh, R., Gupta, A., Srinivasan, G., & Rathore, L. S. (2006). Impact of climate change on Indian agriculture: A review. *Climate Change, 78*, 445–478.

Manderschei, R., & Weigel, H. J. (2000). Drought stress effects in wheat are mitigated by atmospheric carbon dioxide enrichment. *Agronomy for Sustainable Development, 27*, 79–87.

Marilley, L., Hartwig, U. A., & Aragno, M. (1999). Influence of an elevated atmospheric CO_2 content on soil and rhizosphere bacterial communities beneath *Lolium perenne* and *Trifolium repens* under field conditions. *Microbiology Ecology, 38*, 39–49.

Matsui, T., Omasa, T., & Horie, T. (1997). High temperature induced spikelet sterility of japonica rice at flowering in relation to air temperature, humidity and wind velocity. *Japanese Journal of Crop Science, 66*, 449–455.

Mauney, J. R., Kimball, B. A., Printer, P. J., La More, R. L., Lewis, K. F., Nagy, J., et al. (1994). Growth and yield of cotton in response to free air carbon enrichment (FACE) environment. *Agricultural and Forest Meteorology, 70*, 49–68.

Mitchell, R. A. C., Lawlor, D. W., Mitchell, V. J., Gibbard, C. L., White, E. M., & Porter, J. R. (1995). Effects of CO_2 concentration and increased temperature on winter-wheat test of ARCHWHEAT1 simulation model. *Plant, Cell and Environment, 18*, 736–748.

Moldestad, A., Fergestad, E. M., Hoel, B., & Skjelvag, A. O. (2011). Effect of temperature variation during grain filling on wheat gluten resistance. *Journal of Cereal Science, 53*, 1–8.

Monteith, J. L. (1965). Evaporation and environment. 19th *Symposia of the Society for Experimental Biology, 19*, 205–234. Cambridge: Cambridge University Press.

Monteith, J. L. (1981). Presidential address to the Royal Meteorological Society. *Quarterly Journal of the Royal Meteorological Society, 107*, 749–774.

Morison, J. I. L. (1987). Plant growth and CO_2 history. *Nature, 327*, 560.

Morison, J. I. L. (1998). Stomatal response to increased CO_2 concentration. *Journal of Experimental Botany, 49*, 443–453.

Morita, S., Shiratsuchi, H., Takanashi, J. I., & Fujita, K. (2004). Effect of high temperature on grain ripening in rice plants: Analysis of the effects of high night and high day temperatures applied to the panicle and other parts of the plant. *Japanese Journal of Crop Science, 73*, 77–83.

Muchow, R. C., Sinclair, T. R., & Bennett, J. M. (1990). Temperature and solar radiation effects on potential maize yield across locations. *Agronomy Journal, 82*, 338–343.

Mulholland, B. J., Craigon, J., Black, C. R., Colls, J. J., Atherton, J., & Landon, G. (1997). Impact of elevated atmospheric CO_2 and O_3 on gas exchange and chlorophyll content in spring wheat (*Triticum aestivum* L.). *Journal of Experimental Botany, 48*, 1853–1863.

Nelson, G. C., Rosegrant, M. W., Koo, J., Robertson, R., Sulse, T., Zhu, T., et al. (2009). *Climate change impact on agriculture and costs of adaptation.* Washington, DC: International Food Research Institute.

Ohe, I., Saitoh, K., & Kuroda, T. (2007). Effects of high temperature on growth, yield and dry matter production of rice grown in the paddy field. *Plant Production Science, 10*, 412–422.

Olesen, J. E., Trnka, M., Kersebaum, K. C., Skjelvag, A. O., Seguin, B., Peltonen-sainio, P., et al. (2011). Impacts and adaptation of European crop production systems to climate change. *European Journal of Agronomy, 34*, 96–112.

O'Neill, E. G. (1994). Response of soil biota to elevated atmospheric carbon dioxide. *Plant and Soil, 165*, 55–65.

Ottman, M. J., Kimball, B. A., Pinter, P. J., Wall, G. W., Venderlip, R. L., Leavitt, S. W., et al. (2001). Elevated CO_2 increases sorghum biomass under drought conditions. *New Phytologist, 150*, 261–273.

Pan, D. (1996). *Soybean response to elevated temperature and CO_2* [PhD dissertation]. Gainesville: University of Florida.

Pandey, B., Patel, H. R., & Patel, V. J. (2007). Impact assessment of climate change on wheat yield in Gujarat using CERES-wheat model. *Journal of Agrometeorology, 92*, 149–157.

Parry, M. L., Rosenzweig, C., Iglesias, A., Livermore, M., & Fischer, C. (2004). Effects of climate change on global food introduction under SRES emissions and socioeconomic scenarios. *Global Environmental Change, 14*, 53–67.

Pendall, E., Bridgham, S., Hanson, P. J., Hungate, B., Kicklighter, D. W., Johnson, D. W., et al. (2004). Below–ground process responses to elevated CO_2 and temperature: A discussion of observations, measurement methods, and models. *New Phytologist, 162*, 311–322.

Peng, S., Huang, J., John, E. S., Rebecca, C. L., Romeo, M. V., Xuhua, Z., et al. (2004). *Rice yields decline with higher night temperature from global warming.* www.pnas.org_cgi_doi_10.1073_pnas.0403720101.

Pettigrew, W. T. (2008). The effect of higher temperature on cotton lint yield and fiber quality. *Crop Science, 45*, 278–285.

Phillips, R. P., Finzi, A. C., & Bernhardt, E. S. (2011). Enhanced root exudation induces microbial feedbacks to N cycling in a pine forest under elevated long-term CO_2 fumigation. *Ecology Letters, 14*, 187–194.

Pinter, P. J., Jr., Kimbal, B. A., Wall, G. W., LaMorie, R. L., Adansen, F., & Hunsaker, D. J. (1996). *Face 1995-1996: Effects of elevated CO_2 and soil nitrogen on growth and yield parameters of spring wheat.* Annual Research Report 1997 (pp. 75–78). Phoenix, AZ: US Water Conservation Laboratory ASDA, ARS.

Pinter, P. J., Jr., Kimbal, B. A., Wall, G. W., LaMorie, R. L., Adansen, F., & Hunsaker, D. J. (1997). *Effects of elevated CO_2 and soil nitrogen on final grain yields of spring wheat.* Annual Research Report 1997 (pp. 71–74). Phoenix, AZ: US Water Conservation Laboratory, ASDA, ARS.

Porter, J. R., & Gawith, M. (1999). Temperatures and the growth and development of wheat: A review. *European Journal of Agronomy, 10*, 23–36.

Porter, H., & Navas, M. L. (2003). Plant growth and competition at elevated CO_2: On winners, losers and functional groups. *New Phytologist, 157*, 175–198.

Porter, J. R., & Semenov, M. A. (2005). Crop responses to climatic variation. *Philosophical Transactions of the Royal Society, B: Biological Sciences, 360*, 2021–2035.

Prasad, P. V. V., Boote, K. J., Allen, L. H., Jr., Sheehy, J. E., & Thomas, J. M. G. (2006). Species, ecotype and cultivar differences in spikelet fertility and harvest index of rice in response to high temperature stress. *Field Crops Research, 95*, 398–411.

Prasad, P. V. V., Boote, K. J., Allen, L. H., Jr., & Thomas, J. M. G. (2002). Effects of elevated temperature and carbon dioxide on seed set and yield of kidney beans (*Phaseolus vulgaris* L.). *Global Change Biology, 8*, 710–721.

Prasad, P. V. V., Boote, K. J., Allen, L. H., Jr., & Thomas, J. M. G. (2003). Supra-optimal temperatures are detrimental to peanut (*Arachishypogaea* L.) reproductive process and yield at ambient and elevated carbon dioxide. *Global Change Biology, 9*, 1775–1787.

Prior, S. A., Reunion, G. B., Tolbert, H. A., & Reeves, D. W. (2005). Elevated atmospheric CO_2 effects on biomass production and soil carbon in conventional and conservation cropping systems. *Global Change Biology, 11*, 657–665.

Prior, S. A., Torbert, H. A., Runion, G. B., & Rogers, H. H. (2003). Implications of elevated CO_2 induced changes in agroecosystem productivity. *Journal of Crop Production, 8*, 217–244.

Prior, S. A., Torbert, H. A., Runion, G. B., Rogers, H. H., Wood, C. W., Kimball, B. A., et al. (1997). Free air carbon dioxide enrichment of wheat: Soil carbon and nitrogen dynamics. *Journal of Environmental Quality, 26*, 1161–1166.

Prior, S. A., Watts, D. B., Arriaga, F. J., Runion, G. B., Torbert, H. A., & Rogers, H. H. (2010). Influence of elevated CO_2 and tillage practice on precipitation simulation. In *Conservation agriculture impacts—Local and global. Proceedings, 32nd Southern Conservation Agricultural Systems Conference, Jackson, TN* (pp. 83–88).

Pushpalatha, P., Sharma-natu, P., & Ghildiyal, M. C. (2008). Photosynthetic response of wheat cultivar to long-term exposure to elevated temperature. *Photosynthetica, 46,* 552–556.

Rachmilevitch, S., Cousins, A. B., & Bloom, A. J. (2004). Nitrate assimilation on plant shoots depends upon photorespiration. *Proceedings of National Academy Science, USA, 101,* 11506–11510.

Rana, R. S., Bhagat, R. M., Kalia, V., & Lal, H. (2010). Impact of climate change on shift of apple belt in Himachal Pradesh. In P. Shushma, S. R. Shibendu, & S. P. Jai (Eds.), *Impact of climate change on agriculture. Workshop Proceedings, ISPRS.WGVIII–8/W3, Space Application Centre, Ahmadabad* (pp. 131–137).

Randall, P. J., & Moss, H. J. (1990). Some effects of temperature regime during grain filling on wheat quality. *Australian Journal of Agricultural Research, 41,* 603–617.

Randlett, D. L., Zak, D. R., Pregitzer, K. S., & Curtis, P. S. (1996). Elevated atmospheric carbon dioxide and leaf litter chemistry: Influences on microbial respiration and net nitrogen mineralization. *Soil Science Society of America Journal, 60,* 1571–1577.

Reddy, K. R., Doma, P. R., Means, L. O., Hodges, H. F., Richardson, A. G., & Kakani, V. J. (2002). Simulating the impact of climate change on cotton production in Mississippi Delta. *Climate Research, 22,* 271–281.

Reddy, K. R., Hodges, H. F., & Kimball, B. A. (2000). Crop ecosystem responses to global change. In K. R. Reddy & H. F. Hodges (Eds.), *Climate change and global crop productivity* (pp. 161–187). Wallingford: CABI Publication.

Reddy, K. R., Hodges, H. F., McKinion, J. M., & Wall, G. W. (1992). Temperature effects on Pima cotton growth and development. *Agronomy Journal, 84,* 237–243.

Reddy, K. R., Koti, S., Davidonis, G. H., & Reddy, V. R. (2004). Interactive effects of carbon dioxide and nitrogen nutrition on cotton growth, development, yield, and fiber quality. *Agronomy Journal, 96,* 1148–1157.

Reddy, V. R., Reddy, K. R., Acock, M. C., & Trent, A. (1994). Carbon dioxide enrichment and temperature effects on cotton leaf initiation and development. *Biotronics, 23,* 59–74.

Reddy, A. R., Reddy, K. R., & Hodes, H. F. (1998). Interactive effects of elevated carbon dioxide and growth temperature on photosynthesis of cotton leaves. *Plant Growth Regulation, 22,* 1–8.

Reidsma, P., Ewwert, F., Lansink, A. O., & Leemans, R. (2010). Adaptation to climate change and climate variability in European agriculture; the importance of farm level responses. *European Journal of Agronomy, 32,* 91–102.

Resurreccion, A. P., Hara, T., Juliano, B. O., & Yoshida, S. (1977). Effect of temperature during ripening on grain quality of rice. *Soil Science & Plant Nutrition, 23,* 109–112.

Ritchie, J. T., & NeSmith, S. (1991). Temperature and crop development. In R. J. Hanks & J. T. Ritchie (Eds.), *American Society of Agronomy: Vol. 31. Modeling Plant and Soil Systems* (pp. 5–29).

Rogers, A., Ainsworth, E., & Leakey Andrew, D. B. (2009). Will elevated carbon dioxide concentration amplify the benefits of nitrogen fixation in legumes? *Plant Physiology, 151,* 1009–1016.

Rogers, G. S., Milham, P. J., Gillings, M., & Conroy, J. P. (1996). Sink strength may be the key to growth and nitrogen responses in N-deficient wheat at elevated CO_2. *Australian Journal of Plant Physiology, 23,* 253–264.

Rosenzwig, C., Parry, M. L., Fischer, G., & Frohberg, K. (1993). *Climate change and world food supply.* Research report No 3, Environmental Change Unit (pp. 28)Oxford: Oxford University.

Roudier, P., Sultan, B., Quirion, P., & Berg, A. (2011). The impact of future climate change on West African crop yields: What does the recent literature says? *Global Environmental Change, 21,* 1073–1083.

Runge, E. C. A. (1968). Effect of rainfall and temperature interaction during the growing season of corn. *Agronomy Journal, 60,* 503–507.

Sahoo, S. K. (1999). *Simulating growth and yield of maize in different agroclimatic regions* [MSc thesis]. New Delhi: Indian Agricultural Research Institute.

Samarakoon, A. B., & Gifford, R. M. (1995). Soil water content under plants at high CO_2 concentration and interaction with the direct CO_2 effects: A species comparison. *Journal of Biogeography, 22,* 193–202.

Samarakoon, A. B., & Gifford, R. M. (1996). Elevated CO_2 effects on water use and growth of maize in wet and drying. *Australian Journal of Plant Physiology, 23,* 53–62.

Samarakoon, A. B., Muller, W. J., & Gifford, R. M. (1995). Transpiration and leaf area under elevated CO_2-effect of soil water status and genotype in wheat. *Australian Journal of Plant Physiology, 22,* 33–44.

Sankaranaryanan, K., Praharaj, C. S., Nalayani, P., Bandyopadhyay, K. K., & Gopalakrishanan, N. (2010). Climate change and its effect on cotton (*Gosssypium* sp.). *Indian Journal of Agricultural Science, 80,* 561–575.

Saseendran, A. S., Singh, K. K., Rathore, L. S., Singh, S. V., & Sinha, S. K. (2000). Effects of climate change on rice production in the tropical humid climate of Kerala, India. *Climate Change, 44,* 495–514.

Schortemeyer, M., Hartwig, U. A., Hendrey, G. R., & Sadowsky, M. J. (1996). Microbial community changes in the rhizospheres of white clover and perennial ryegrass exposed to free air carbon dioxide enrichment (FACE). *Soil Biology and Biochemistry, 28,* 1717–1724.

Schutz, M., & Fangmeier, A. (2001). Growth and yield responses of spring wheat (*Triticum aestivum* L. cv. Minaret) to elevated CO_2 and water limitation. *Environmental Pollution, 114,* 187–194.

Serbin, P. S., & Kucharik, C. J. (2009). Spatiotemporal mapping of temperature and precipitation for the development of the multidecadal climatic dataset for Wisconsin. *Journal of Applied Meteorology and Climatology, 48,* 742–757.

Sinclair, T. R., Pinter, P. J., Jr., Kimball, B. A., Adamsen, F. J., LaMore, R. L., Wall, G. W., et al. (2000). Leaf nitrogen concentration of wheat subjected to elevated CO_2 and water or N deficits. *Agriculture, Ecosystems and Environment, 79,* 53–60.

Singletary, G. W., Banisadr, R., & Keeling, P. L. (1994). Heat stress during grain filling in maize: Effects on carbohydrate storage and metabolism. *Australian Journal of Plant Physiology, 21,* 829–841.

Southworth, J., Pfeifer, R. A., Habeck, M., Randolph, J. C., Doering, O. C., Johnston, J. J., et al. (2002a). Changes in soyabean yields in mid-western United States as a result of future changes in climate variability, and CO_2 fertilization. *Climate Change, 53,* 447–475.

Southworth, J., Pfeifer, R. A., Habeck, M., Randolph, J. C., Doering, O. C., & Rao, G. D. (2002b). Sensitivity of winter wheat in mid-western United States as a result of future changes in climate variability, and CO_2 fertilization. *Climate Change, 22,* 73–86.

Southworth, J., Randolph, J. C., Habeck, M., Doering, O. C., Pfeifer, R. A., Rao, D., et al. (2000). Consequences of future climate change and changing climate variability on maize yields in mid-western United States. *Agriculture, Ecosystems and Environment, 82,* 139–158.

Stake, T., & Hayase, H. (1970). Male sterility caused by cooling treatment at young microspore stage in rice plants.5. Estimation of pollen development stage and most sensitive stages of coolness. *Proceedings of crop Sciences Society of Japan, 39,* 468–473.

Stoate, C., Baldi, A., Beja, P., Boatman, N. D., Herzon, I., Van Doorn, A., et al. (2009). Ecological impacts of early 21st century agricultural change in Europe—A review. *Journal of Environmental Management, 91,* 22–46.

Subash, N., & Ram Mohan, H. S. (2012). Evaluation of the impact of climatic trends and variability in rice-wheat system productivity using cropping system model DSSAT over the Indo-Gangetic plains of India. *Agricultural and Forest Meteorology, 164,* 71–81.

Taneva, L., Pippen, J. S., Schlesinger, W. H., & Gonzalez-Meler, M. A. (2006). The turnover of the carbon pools contributing to soil CO_2 and soil respiration in a temperate forest exposed to elevated CO_2 concentration. *Global Change Biology, 12*, 983–994.

Tarnawski, S., Hamelin, J., Jossi, M., Aragno, M., & Fromin, N. (2006). Phenotypic structure of Pseudomonas populations is altered under elevated pCO_2 in the rhizosphere of perennial grasses. *Soil Biology and Biochemistry, 38*, 1193–1201.

Tashiro, T., & Wardlaw, I. F. (1989). A comparison of the effect of high-temperature on grain development in wheat and rice. *Annals of Botany, 64*, 59–65.

Tashiro, T., & Wardlaw, I. F. (1990). The response of high temperature shock and humidity changes prior to and during the early stages of grain development in wheat. *Australian Journal of Plant Physiology, 17*, 551–561.

Taub, D., Miller, B., & Allen, H. (2008). Effects of elevated CO2 on the protein concentration of food crops: A meta-analysis. *Global Change Biology, 14*, 565–575.

Taub, D. R., & Wang, X. Z. (2008). Why are nitrogen concentrations in plant tissues lower under elevated CO2? A critical examination of the hypotheses. *Journal of Integrative Plant Biology, 50*, 1365–1374.

Tester, R. F., Morrison, W. R., Ellis, R. H., Piggot, J. R., Batts, G. R., Wheeler, T. R., et al. (1995). Effect of elevated growth temperatures and carbon dioxide levels on some physicochemical properties of wheat starch. *Journal of Cereal Science, 22*, 63–71.

Tiwari, S., & Agrawal, M. (2006). Evaluation of ambient air pollution impact on carrot plants at a suburban site using open top chamber. *Environmental Monitoring and Assessment, 119*, 15–30.

Torbert, H. A., Prior, S. A., Rogers, H. H., & Runion, G. B. (2004). Elevated atmospheric CO_2 effects on N fertilization in grain sorghum and soybean. *Field Crops Research, 88*, 57–67.

Tubiello, F. N., Armthor, J. S., Boote, K. J., Donatelli, M., Easterling, W., Fischer, G., et al. (2007). Crop response to elevated CO_2 and world food supply—A comment on 'food for thought' by Long et al., Science. 312: 1918–1921, 2006. *European Journal of Agronomy, 26*, 215–223.

Tubiello, F. N., & Ewert, F. (2002). Simulating the effects of elevated CO_2 on crops: Approaches and applications for climate change. *European Journal of Agronomy, 18*, 57–74.

Valle, R., Mishoe, J. W., Jones, J. W., & Allen, L. H., Jr. (1985). Transpiration rate and water use efficiency of soybean leaves adapted to different CO_2 environments. *Crop Science, 25*, 477–482.

Vanaja, M., Raghuram, R., Lakshmi, N. J., Maheshvari, M., Vagheera, P., Ramkumar, P., et al. (2007). Effect of elevated CO_2 concentration on growth and yield of black gram (*Vigna mungo* L. Hepper): A rain-fed pulse crop. *Plant, Soil and Environment, 53*, 81–88.

Vara Prasad, P. V., Craufurd, P. Q., Summerfield, R. J., & Wheeler, T. R. (2000). Effects of short-episodes of heat stress on flower production and fruit-set of groundnut (*Arachis hypogaea*L.). *Journal of Experimental Botany, 51*, 777–784.

Vashisht, B. B., & Jalota, S. K. (2014). In *Wheat yield response to temperature variability in central Punjab. Abstracts of ISTS-IUFRO conference on sustainable resource managemet for climate change mitigation and social security* (pp. 11–12).

Vashisht, B. B., Mulla, D. J., Jalota, S. K., Kaur, S., Kaur, H., & Singh, S. (2013). Productivity of rain-fed wheat as affected by climate change scenario in northeastern Punjab, India. *Regional Environmental Change, 13*, 989–998.

Vashisht, B. B., Nigon, T., Mulla, D. J., Rosen, C., Xu, H., Twine, T., et al. (2015). Adaptation of water and nitrogen management to future climates for sustaining potato yield in Minnesota: Field and simulation study. *Agricultural Water Management, 152*, 198–206.

Von Caemmerer, S., & Furbank, R. T. (2003). The C4 pathway: An efficient CO_2 pump. *Photosynthesis Research, 77*, 191–207.

Wand, S. J. E., Midgley, G. F., Jones, M. H., & Cuttis, P. S. (1999). Response of wild C4 and C3 grasses (Poaceae) species to elevated atmospheric CO_2 concentration, a meta-analysis test of current theories and perceptions. *Global Change Biology, 5*, 23–741.

Weigel, H. J., Manderscheid, R., Jager, H. J., & Mejer, G. J. (1994). Effects of season long CO_2 enrichment on cereals. I. Growth performance and yield. *Agroecosystems and Environment, 48*, 231–240.

Weightman, R. M., Miller, S., Alava, J., Foulkes, M. J., Fish, L., & Snape, J. W. (2008). Effects of drought and the presence of the IBL/IRs translocation on grain vitreosity, hardness and protein content in winter wheat. *Journal of Cereal Science, 47*, 457–468.

Welch, J. A., Vincent, J. R., Auffhammer, M., Moya, P. F., Dobermann, A., & Dawe, D. (2010). Rice yields in tropical/sub-tropical Asia exhibit large but opposing sensitivities to minimum and maximum temperatures. *Proceedings of National AcademyScience, USA, 107*, 14562–14567.

Wheeler, T. R., Craufurd, P. Q., Ellis, R. H., Porter, J. R., & Vara Prasad, P. V. (2000). Temperature variability and the yield of annual crops. *Agriculture, Ecosystems and Environment, 82*, 159–167.

Williams, M. A., Rice, C. W., & Owensby, C. E. (2000). Carbon dynamics and microbial activity in tall grass prairie exposed to elevated CO_2 for 8 years. *Plant and Soil, 227*, 127–137.

Wolf, J., & Van Oijen, M. (2003). Model simulation of effects of changes in climate and atmospheric CO_2 and O_3 on tuber yield potential of potato (cv. Bintje) in the European Union. *Agriculture, Ecosystems and Environment, 94*, 141–157.

Wong, S. C. (1980). Effects of elevated partial pressures of CO_2 on rate of CO_2 assimilation and water use efficiency in plants. In G. I. Pearman (Ed.), *Carbon dioxide and climate: Australian research* (pp. 159–166). Canberra, Australia: Australian Academy of Science.

Xiao, G., Liu, W., Xu, Q., Sun, Z., & Wang, J. (2005). Effects of temperature increase and elevated CO_2 concentration, with supplemental irrigation on the yield of rain-fed spring wheat in semi-arid region of China. *Agricultural Water Management, 74*, 243–255.

Xiao, G., Zang, Q., Yao, Y., Yang, S., Wang, R., Xiong, Y., et al. (2007). Effects of temperature increase on water use and crop yields in a pea-spring wheat-potato rotation. *Agricultural Water Management, 91*, 86–89.

Yamakawa, H., Hirose, T., Kuroda, M., & Yamaguchi, T. (2007). Comprehensive expression profiling of rice grain filling-related genes under high temperature using DNA microarray. *Plant Physiology, 144*, 258–277.

Yoshida, S., Satake, T., & Mackill, D. S. (1981). Heat temperature stress in rice. IRRI research paper series. IRRI, Manila, Philipines, Vol. 67, (pp. 1–15).

Yoshimoto, M., Oue, H., & Kobayashi, K. (2005). Responses of energy balance, evapotranspiration, and water use efficiency of canopies to free-air CO_2 enrichment. *Agriculture and Forest Meteorology, 133*, 226–246.

Zhang, B., liu, W., Chang, S. X., & Anyia, A. O. (2010). Water deficit and high temperature water use efficiency and arbinoxylan concentration in spring wheat. *Journal of Cereal Science, 52*, 263–269.

Zhou, X. D., Wang, F. T., & Zhu, Q. J. (2002). Numerical simulation study on the effects of increasing CO_2 concentration on winter wheat. *Acta Meterologica Sinica, 60*, 53–59.

Zhu, B., & Cheng, W. X. (2011). Rhizosphere priming effect increases the temperature sensitivity of soil organic matter decomposition. *Global Change Biology, 17*, 2172–2183.

Ziska, L. H., Reves, J. B., & Blank, B. (2005). The impact of recent increase in atmospheric CO_2 on biomass production and vegetative retention of cheat grass (*Bromus tectorum*): Implications for fire disturbance. *Global Change Biology, 11*, 1325–1332.

Ziska, L. H., & Teasdale, J. R. (2000). Sustained growth and increased tolerance to glyphosate observed in a C3 perennial weed, quackgrass (*Elytrigia repens*), grown at elevated carbon dioxide. *Australian Journal of Plant Physiology, 27*, 159–166.

FURTHER READING

https://buythetruth.Wordpress.com/2009/06/24/met-office-fraudcast.
https://co2now.org/Current-CO2/CO2-Now/global-carbon-emissions.html.
Miller, P., Lanier, W., & Brandt, S. (2001). Using growing degree days to predict plant stages. Montana State University Extension Service MT200103 AG 7/2001.

CHAPTER FOUR

Climate Change and Groundwater

Contents

4.1 INTRODUCTION

Water resources on the earth are in abundance; approximately two-third areas of the earth are covered with water. Of the total water, 97.0% is stored in the oceans that are salty, and 3.0% is freshwater. The freshwater includes ice in glaciers and polar ice sheets (69.4%); water stored in lakes, rivers, and wetlands and vapor in the atmosphere (0.3%); soil moisture (0.9%); and groundwater, the water stored underground in fully saturated soils and geologic formations (29.4%). But the use of the total water is restricted due to nonsynchronization of the availability and demand at temporal and spatial scales and poor quality. With ever-increasing global population, the demand of clean water for agricultural, household, recreational, industrial, and environmental use is rising. In case of nonavailability of surface water, the users mainly rely on groundwater for meeting the spatial and temporal water demands, which has its own importance in the hydrologic cycle. Globally, groundwater is the source of one-third of all freshwater withdrawals and supplies 36%, 42%, and 27% of the water for domestic, agricultural, and industrial purposes, respectively (Doll et al., 2012). Natural groundwater discharges help sustain base flow to rivers, lakes, and wetlands during periods of low or no precipitation (P_{cp}).

Climate change being an ongoing phenomenon inevitably brings about numerous environmental changes, that is, increased CO_2, temperature, and P_{cp}. These changes not only affect the crop productivity and root-zone water balance (discussed in Chapter 3) but also impact the total hydrologic cycle. In many cases, the data show that the hydrologic cycle has already been influenced by climate change, though the impact is greatly influenced by anthropogenic activity (Bernstein et al., 2008). In the future, the continuous increase in climate warming, variability, extremes, and other changes may influence groundwater systems and cause severe impact on water resources. Despite understanding the groundwater's significance, most of the climate change impact studies are concentrated on surface water resources; however, a few are on groundwater (Chen, Li, Zhang, & Ni, 2014; Dragoni & Sukhija, 2008; Taylor et al., 2013). Climate change may impact the groundwater directly through replenishment by recharge and indirectly through changes in its use. These impacts can be modified by human activity mainly by land use changes (LUC). Such studies on the groundwater response to climate change together with human activities have started to come up recently (Green et al., 2011; Taylor et al., 2013).

This chapter aims at understanding the impact of increased CO_2, temperature, and P_{cp} due to climate change on hydrologic cycle components, such as evapotranspiration (ET), runoff, recharge and discharge, and quality change of groundwater. Methodologies to measure recharge, discharge, and fluxes across boundaries are also talked about. It comprises information on the effect of land use and land cover, vegetation, and geology of the aquifer on recharge. Emission of gases due to energy is expended on groundwater pumping, groundwater estimation by coupling climate, soil-water-crop, and groundwater models through geographic information system (GIS). The management interventions of enhancing the surface-water supply and reducing groundwater draft to protect the groundwater resources from the impact of climate change are also discussed.

4.2 HYDROLOGICAL CYCLE

The hydrologic cycle (Fig. 4.1) represents the continuous movement of water between the atmosphere, the Earth's surface (glaciers, snowpack, streams, wetlands, and oceans), soils, and geologic formations. The hydrologic cycle is driven by solar energy, which heats the Earth's surface and ocean and causes water loss to the atmosphere through evaporation, sublimation, and transpiration. Water is transported from the atmosphere back to the Earth's surface as P_{cp} or snow. Water from P_{cp} enters the soil through infiltration (*the process of entry of water on the soil surface and movement in subsurface*). The water exceeding infiltration capacity of the soil goes as runoff depending upon the surface conditions, soil profile, and P_{cp} intensity. Tilled soil surfaces and coarse-textured soil profiles have more infiltration rate and less runoff than compacted and fine-textured soils. Infiltration is more when P_{cp} is in light showers. The infiltrated water gets accumulated in unsaturated zone

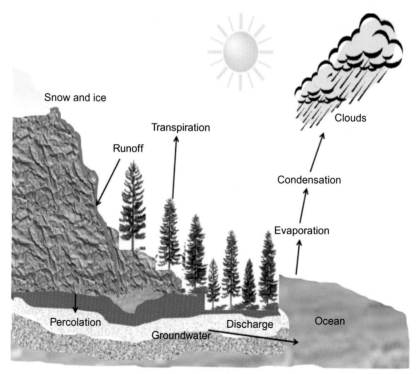

Fig. 4.1 Hydrologic cycle.

of the soil and goes on moving in all directions, that is, downward, upward, or sideways depending upon its water retention capacity and potential gradient. In many unsaturated zone studies, some terms such as percolation (*water migration down through the soil profile*) or deep drainage (*the water that has moved across the root zones*) have been used to describe water movement within the soil.

In vegetation-covered soils, a part of the infiltrated water is taken by plants through their roots from the root–zone soil and is lost to the atmosphere by transpiration; a part is lost from soil surface as evaporation, and the rest leaches to the vadose zone (*the zone beyond root zone to groundwater level*). Water in the vadose zone moves in response to gradient of total soil water potential (gravitational plus pressure potentials) and water transmission properties of the soil and reaches to groundwater. The water that leaves the vadose zone and contributes to groundwater is called recharge. In other words, the groundwater recharge is the residual flux of water added to the saturated zone resulting from the evaporative, transpirative, and runoff losses of the P_{cp} and snow that causes groundwater accretion. The groundwater gets lost into the ocean through discharge and to the atmosphere through pumping for domestic, agricultural, and industrial uses on the Earth's surface. The hydrologic cycle clearly signifies that groundwater evolution is the resultant of the two main processes, recharge and discharge, which need proper understanding.

4.2.1 Recharge

Recharge from P_{cp} can take place by diffuse infiltration, a preferential pathway and/or through surface streams and lakes. It follows the three routes of water, that is, direct, localized, and indirect. The direct recharge occurs through the unsaturated zone when water from P_{cp}, in excess of soil moisture deficit and ET, percolates in vertical direction directly to the groundwater system. It (direct recharge) is a diffuse process and occurs fairly uniformly over large areas beneath the point of the impact of the P_{cp}. Localized recharge results from horizontal movement and concentration of water into joints, rivulets and depressions. These focused points are too frequent to be mappable and measureable. The focused recharge beneath ephemeral surface-water bodies is mostly generated in semiarid environment when P_{cp} is statistically extreme (heavy). Indirect recharge involves concentration of water usually in rivers, whether ephemeral or perennial via leakage from surface waters (i.e., ephemeral streams, wetlands, or lakes). It results from the changes in groundwater use and can be modified by human activity such as land use change and is different from the localized recharge as the water courses are significantly large to be mapped, counted, and possibly gauged. These distinctions are made based on the processes and of practicality and cannot be adhered to. For example, preferential pathway of recharge such as fissure and root holes leads to direct recharge. Actual recharge (estimated from groundwater studies) may be less than the potential recharge, that is, *maximum water available for to become recharge*. Potential recharge results in an overestimate of actual recharge because of reservoir storage and subsequent ET, development of perched aquifers, and inability of the aquifer to accept recharge because of a shallow water table or low water transmission capacity. The recharge is affected by a number of factors, and its magnitude is highly dependent on the prevailing climate, land cover, and underlying geology. Climate and land cover largely determine P_{cp} and ET, whereas the underlying soil and geology dictate whether a surplus water (P_{cp} minus ET) can be transmitted and stored in the subsurface. Recharge is increased with P_{cp} and vice versa. But the contribution of P_{cp} to change the groundwater level through recharge gets started only after a part of P_{cp} is utilized to satisfy the soil water deficit (field capacity minus soil water content) in the vadose zone and a part is consumed in the perched water table. In other words, recharge occurs only in the season when P_{cp} (input) exceeds ET (output). For example, in Indian Punjab under the existing environmental conditions of the water supply and demand, rise in groundwater level through recharge occurs only during the rainy season when monsoon P_{cp} exceeds 515 mm (Fig. 4.2).

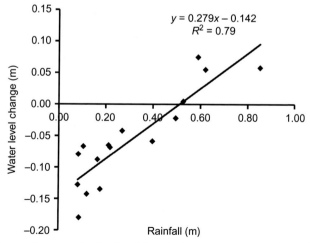

Fig. 4.2 Change in groundwater level in relation to monsoon rain under present surface-water supply and water demand.

The recharge is more in *Kharif* (local name of a cropping season when P_{cp} exceeded ET, period is from mid-June to mid-October) than *Rabi* (local name of a cropping season when ET exceeds P_{cp}, period is from mid-October to mid-June). In *Rabi* season, water level recedes due to more extraction of groundwater than the P_{cp} to meet the ET demand (Figs. 4.3 and 4.4). Figs. 4.3 and 4.4 and the results of Rosenberg et al. (1999) clearly point out that under irrigated agriculture system, recharge is likely to increase in the seasons when P_{cp} is above normal, and water demand is less due to lower solar radiation and high humidity resulting from cloudy sky.

As P_{cp} is location-specific, so is the recharge. There is general agreement that many areas of currently high P_{cp} are expected to experience P_{cp} increases, whereas many of the areas at present with low P_{cp} and high evaporation, now suffering water scarcity, are expected to have P_{cp} decreases (Bernstein et al., 2008).

In the areas with increased P_{cp}, recharge will increase, while in the areas already suffering from water scarcity and in those where the P_{cp} will decrease and the climate will get drier, recharge will decrease. In areas where permafrost thaws, groundwater recharge is more, and the magnitude of recharge is modified by latitudes and elevations, which decide snow accumulation and its melting time. At high latitudes and elevations, less snow accumulation and earlier melting of snow due to global warming change the spatial and temporal distribution of snow and ice, which tend to reduce the seasonal duration and magnitude of recharge. In mountain valleys, due to warming, aquifers show (i) shifts in the timing and magnitude of peak groundwater levels due to an earlier spring melt and (ii) low groundwater levels associated

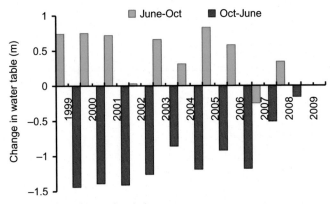

Fig. 4.3 Seasonal groundwater level changes.

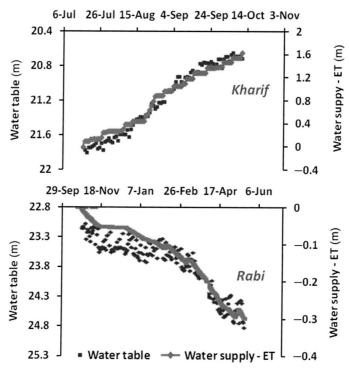

Fig. 4.4 Periodic water supply (rain + surface water) minus ET and groundwater change during *Kharif* (rice) and *Rabi* (wheat) seasons in 2008.

lower base flow with longer periods. At higher altitudes where glaciers are water resources, glaciation conditions determine the recharge. Unstable glaciers generate more runoff and result in less recharge than stable (Akhtar, Ahmed, & Booij, 2008).

In the glaciated watersheds of the Himalayas, the impacts of large reductions in glacial mass and increased evaporation on groundwater recharge are projected to be offset by a rise in P_{cp}. In areas of seasonal or perennial ground frost, increased recharge is expected even though the absolute snow volume decreases. In addition, different forms of water like P_{cp} and snow, atmospheric CO_2, and temperature also affect magnitude of groundwater recharge by controlling ET. The dependence of ET on stomatal closure, extension of foliage size and crop duration by CO_2, and shortening crop duration along with increasing the vapor pressure gradient between leaf and atmosphere by increased temperature has already been explained in Chapter 3.

Recharge is also influenced by land use and land cover (LULC) and depth of groundwater level; however, the recharge responses exceed greatly those resulting from climate variability including severe droughts. Ground-water recharge is generally lower in cropped and forested areas than in non-cropped and nonforested (Scanlon et al., 2006). In cropped and forested areas, the major part of the surface-water input (P_{cp} + irrigation) is lost as ET, leaving a fraction of water for recharge. There is another view that vegetation can increase recharge by (i) facilitating infiltration and reducing overland flow; (ii) creating surface storing opportunities; (iii) creating macropores and channels in the soil profile by the plant root systems; and (iv) transferring the water downward due to differences in moisture potentials. Thus, the effect of vegetation on recharge can be in both ways positive and negative. From a very thorough review of 98 groundwater recharge studies in arid and semiarid regions, Scanlon et al. (2006) concluded that LULC is the factor that most regularly controls recharge and affects the water availability for human beings and land degradation. For example, in semiarid areas of Spain, human water availability was decreased due to less recharge with increased vegetative land cover (Bellot, Bonet, Sanchez, & Chirino, 2001). On the contrary, substantial water logging, salinity problems, and flushing of the large amounts of chlorides arose in Australia due to increased recharge rates resulted from clearing of native vegetation and replacing it with shallow-rooted annual crops and pastures (Scanlon et al., 2006; Walker, Zhang, Ellis, Hatton, & Petheram, 2002). In South India while assessing the effect of land use and soil type on groundwater budgets at the subwatershed scale, Anuraga, Ruiz, Kumar, Sekhar, and Leijnse (2006) concluded that (i) the effect of soil type is more prominent than that of land use and (ii) the groundwater balance is more dependent on the amount of withdrawal than on the recharge rate. In addition to climate, LULC, and soil type, underlying geology also affects the recharge. Recharge is decreased if the vadose zone is constituted by layer of low permeability or there are hard or claypans. With increasing urbanization, groundwater recharge is decreased profoundly due to increasing the amount of impervious areas, which alters the land surface from its natural state (Lerner, 2002; Lerner, Issar, & Simmers, 1990). The increased surface runoff due to imperviousness has been simulated in Belgium using MODFLOW (Batelaan, De Smedt, & Triest, 2003). Changes in land use have led to changes in the recharge pattern in rural area of southern England but had no impact on the overall catchment (Finch, 2001). A diminution of agricultural areas due to increased concrete area reduces recharge (Dams, Woldeamlak, & Batelaan, 2007). Recharge to an aquifer depends on the groundwater level too; lower positions of groundwater normally increase recharge by increasing the capture zone.

4.2.2 Discharge

Groundwater discharge is another key element in the water cycle, which includes loss of water from the aquifers to surface water and atmosphere as base flow into streams, springs, wetlands, and oceans and extraction for human needs. Discharge is driven by relative head levels between groundwater and surface water, which are changed with pumping of groundwater for agriculture and increased population. Groundwater discharge continues till groundwater level remains higher than surface-water level and vice versa. Due to this reason, base flow is maintained in permanent streams during periods of extended drought in semiarid and arid regions. In temperate areas where higher winter recharge is projected (e.g., United Kingdom), it is likely that some watersheds could sustain higher base flows during summer even if summers become warmer and drier (Acreman, 2000). The discharge in wet season accompanied by runoff causes flooding, while under dry climate in all seasons, recharge decreases and results in decline in water level, which affects local aquatic life adversely due to less base flow maintained by groundwater discharge (Woldeamlak, Batelaan, & Smedt, 2007). A review paper, presenting the methods for quantifying submarine groundwater discharge, indicates that the process is essentially ubiquitous in coastal areas (Burnett et al., 2006).

4.2.3 Groundwater Storage

Groundwater storage is the difference between recharge and discharge over the time frames that these processes occur, ranging from days to thousands of years. Changes to both groundwater and surface-water levels may ultimately alter the interaction between groundwater and surface water and the interaction between natural and societal water supply and demand (Hanson et al., 2012). Groundwater storage is affected by the intrinsic properties of the aquifer (storage capacity, transmission capacity, and aquifer geometry) controlled by its size and type. An aquifer receiving recharge from extensive catchment areas is insensitive to short-term climatic variability, while shallow unconfined aquifers are more responsive to smaller-scale climate variability. Deeper aquifers react with delay to large-scale climate change only, not to short-term climate variability. Shallow groundwater systems (especially unconsolidated sediment or fractured bedrock aquifers) are more responsive to smaller-scale climate variability (Kundzewicz & Doell, 2009). Groundwater storage is also influenced by climate change depending upon whether groundwater is renewable or not (contemporary recharge) or comprises a fossil resource (Hanson, Newhouse, & Dettinger, 2004).

BOX 4.1

1. Confined aquifers with upper impermeable layers where recharge only occurs from precipitation where the water-bearing formations outcrop at land surface.
2. Unconfined (phreatic) aquifers in wet regions where rainfall is high and evapotranspiration is low. These aquifers are highly renewable because precipitation exceeds evapotranspiration throughout much of year and are not expected to face substantial threats to climate change. Nonrenewable groundwater is vulnerable to the indirect effects of increased abstraction by humans to meet current water requirements and future water demand under a changing climate.
3. Unconfined aquifers in semiarid and arid regions that are likely to have shifting annual balances between precipitation and evapotranspiration and a general drying trend under most climate change forecasts. It is suggested that recharge may be less to these aquifers, resulting in less groundwater availability but an increase in demand from growing population and less reliable surface-water resources.
4. Coastal aquifers vulnerable to rising sea levels and salt-water intrusion.

The groundwater storage will be more sensitive to climate if it is renewable. For more clarity, climate change effects on recharge and groundwater resources are divided into four categories (Sharif & Singh, 1999; Box 4.1).

4.3 MEASUREMENT AND ESTIMATION OF GROUNDWATER BALANCE

Groundwater balance of a region is estimated by quantification of all individual inflows to or outflows from a groundwater system as Eq. (4.1):

$$Rr + Rc + Ri + Rt + Si + Ig = \text{ET} + Tp + Se + Og + \Delta S \qquad (4.1)$$

where Rr is recharge from P_{cp}, Rc is recharge from canal seepage, Ri is recharge from field irrigation, Rt is recharge from tanks, Si is influent seepage from rivers, Ig is inflow from other basins, ET is evapotranspiration from groundwater, Tp is draft from groundwater, Se is effluent seepage to rivers, Og is outflow to other basins, and ΔS is change in groundwater storage. Preferably, all elements of the groundwater balance equation are to be computed using independent methods. But it is not always possible to compute all individual components of the groundwater balance equation separately. Sometimes, depending on the problem, some components are to be lumped, and only their net value is accounted for in the equation.

4.3.1 Recharge

Recharge being a more important component of groundwater accretion, its precise measurement or estimation is important to know the change in groundwater storage as influenced by climate change. A number of methods are available for estimating recharge, but it is generally accepted that there is no universally applicable technique for accurately estimating the recharge rate. Actually, recharge processes vary from one place to another, and there is no guarantee that a method developed and used for one locality will give reliable results when used in another. Recharge may occur naturally from P_{cp}, rivers, canals, and lakes, and man-induced through activities of irrigation and urbanization (roof top rainwater harvesting). Recharge rates from P_{cp} alone have been estimated by using empirical relationships. But such estimates do not account for the principal hydraulic processes (infiltration, drainage, percolation, etc.) and spatial variability (the presence and permeability of soil layers) actually affecting recharge. The estimation of recharge is further complicated by irrigation, which may simultaneously remove water from focused recharge sources while creating new sources of diffuse recharge. In irrigated system, return flow also interferes to recharge because it is a part of groundwater withdrawal that joins back to the groundwater. Estimating groundwater recharge in arid and semiarid regions is much more difficult, since in such areas the recharge is generally low compared with the average annual P_{cp} or evapotranspiration and thus difficult to determine precisely. The recharge is measured by different methods in different situations; therefore, it is necessary to identify the probable flow mechanisms and the important features influencing the recharge in a locality before deciding on the recharge method to use. This necessitates the understanding of methodologies of recharge measurement/estimation, their specialty and limitations, and factors affecting it.

The techniques for estimating recharge are subdivided into various types, on the basis of the hydrologic sources or zones, from which the data are obtained. There are three such hydrologic zones, namely, surface-water zone, unsaturated zone, and saturated zone. Surface-water bodies often form localized recharge sources, and recharge is estimated using surface water gaining and losing data. There is a gain in water in surface-water bodies in humid regions as groundwater is discharged to streams and lakes. In contrast, there is a loss of water in surface-water bodies in arid regions as the surface-water and groundwater systems are often separated by the thick unsaturated sections. Surface-water zone approaches provide estimates of drainage or potential recharge, not actual. The unsaturated zone techniques provide mostly point estimates of drainage or potential recharge. Saturated zone techniques commonly integrate over much larger areas. These provide evidence of actual recharge because water reaches the water table. Within each zone, techniques are generally classified into physical, tracer, and numerical modeling. The physical techniques are channel-water budget, seepage meters, and base-flow discharge in surface-water zone; lysimeters, Darcy's law, and zero-flux plane in unsaturated zone; and water-table fluctuation and Darcy's law in saturated zone. Tracers used are stable isotopes of oxygen (^{18}O) and hydrogen (^{2}H) in surface water; tritium, chloride mass balance, historical tracers (contaminants such as bromide, nitrate, atrazine, and arsenic from industrial and agricultural sources and atmospheric nuclear testing like ^{3}H and ^{36}Cl), and environmental tracers (chloride concentration or chloride mass–balance approach) in unsaturated zone; and ^{3}H concentrations, chlorofluorocarbons (CFC), and radioactive decay of ^{14}C or ^{36}Cl in a confined aquifer in saturated zone. Numerical modeling includes deep percolation model, SWAT model, and water budget equation in surface water and Richard's flow equation in unsaturated and saturated zones. Recently, remote sensing is used to assess the spatial and temporal variations in water fluxes in different components of the water cycle. Any redistribution of water masses in different parts of water cycle may result in time variation of gravity field. The Gravity Recovery and Climate Experiment (GRACE) satellite aimed at observing the gravity field to a high accuracy (c. 1 cm in terms of geoid height and a spatial resolution of 200–300 km) is of such type. This tool has been used successfully to quantify water table decline for the past (Rodell, Velicogna, & Famiglietti, 2009) and for the midcentury (Chen et al., 2014) in northwest India. Detailed description of each of the abovementioned techniques is available in Scanlon, Healy, and Cook (2002) and Dragoni and Sukhija (2008).

4.3.2 Discharge

Discharge is measured/estimated by direct and indirect methods. The direct methods are those in which the volume of water withdrawal is either metered or is calculated by rate–time methods (pumping rate × cumulative pumping time) for a given area. In indirect methods, withdrawals are estimated from (i) theoretical crop water use requirement to a known area under cultivation; (ii) energy consumption; (iii) numerical models; and (iv) recession of water-table levels. In practice, the precise method used to calculate groundwater extraction varies in different regions; for example, in India, groundwater abstraction is estimated using a combination of well census figures, pumping hours, water duties, crop areas, and well yields. The Groundwater Resource Estimation Committee (GWREC, 1997) recommends calculating abstraction by multiplying the average area irrigated by each well by the average annual irrigation depth. In central Spain, discharge is computed as crop water requirements coupling remote sensing techniques with GIS. Introduced stream tracers and continuous measurements of water levels and temperature in shallow piezometers near and in the river also are used to estimate groundwater discharge. Each procedure has some uncertainties; for example, groundwater pumping and actual crop water requirements are unidentical for a given crop because of variation in soil type and potential evapotranspiration controlled by weather and irrigation technologies. Groundwater discharge models have the disadvantage as they need to know all other elements of the groundwater budget precisely both in time and space. Numerical models require comprehensive datasets. Groundwater discharge to streams is measured as water flux (using seepage meters or inferring specific discharge through hydraulic or temperature gradients) and water flow (using current meter). Flux measurements quantify the spatial variability of groundwater discharge and do not necessarily result in mass balance, while flow measurements bring about a lumped discharge value for an entire stream or stream reach and results in mass balance.

Change in water level depth (mm) is measured with piezometers or calculated by dividing the change in storage ($\pm Sg$) with specific yield (*change in water storage per unit change in water level*) as Eq. (4.2):

$$\text{Change in water level} = \pm S_g / \text{Specific yield} \qquad (4.2)$$

Fluxes are measured across boundaries using Darcy's equation, which requires both an accurate physical representation of the system and its differentiation from the adjacent groundwater system and appropriate specified boundary conditions. The absence of well-defined physical boundaries in the near vicinity of zone of interest necessitates a methodology to compute influx and outflux at the boundaries. For that, GIS is an important tool, which could be used along with Darcy's law (Eq. 4.3) to estimate the average flow across a given boundary method:

$$Q = KiA \qquad (4.3)$$

where Q is the flow ($m^3 m^{-2} day^{-1}$), K the average hydraulic conductivity of boundary and adjacent cells ($m\,day^{-1}$), $i = (H_2 - H_1)/L$ the hydraulic gradient (dimensionless), $H_2 - H_1$ the difference in hydraulic heads between two cells (m), and L the distance between two cell nodes (m). A is the area of cross-section (m^2) perpendicular to the flow and equals $B \times W$, where B is the saturated thickness (m) that varies with the change in water-table depth and W is the width of the cell (m). Applying the abovementioned methodology, net flux for a given period can be obtained. GIS also offers facilities for creating profile graphs, extracting values to a point, and creating buffer, interpolation tools, etc. Using sequence of GIS-based tools to calculate gradient across the boundary and then fluxes across such boundaries with Darcy's law from readily available data of water level and transmissivity, Kaur, Aggarwal, Jalota, and Sood (2015) estimated averaged net flux equals to 3.5 $Mm^3 year^{-1}$ (million cubic meter per year) for the years 2001–10 for Ludhiana District of Indian Punjab.

4.4 EFFECT OF CLIMATE CHANGE ON GROUNDWATER

Climate change could affect groundwater sustainability in several ways, including (i) changes in groundwater recharge resulting from seasonal and decadal changes in P_{cp} and temperature; (ii) more severe and longer-lasting droughts; (iii) changes in evapotranspiration due to changes in temperature and vegetation; (iv) possible increased demands for groundwater as a backup source of water supply or for further economical (agricultural) development; and (v) seawater intrusion in low-lying coastal areas due to rising sea levels and reduced groundwater recharge that may lead a deterioration of the groundwater quality there. Climate change effect on groundwater is also modified by the topography, soil profile properties, and geology of the aquifer, as well as anthropogenic activities like change in LULC and population. Multitudes of these factors that concurrently work together make the groundwater recharge highly spatial, which further complicates the problem of defining regional recharge due to involvement of the multicomponent interactions, hydrologic atmospheric processes, and hydrologic boundary conditions. The variations in aquifer recharge not only change the aquifer yield or discharge but also modify the groundwater flow network; for example, gaining streams may suddenly become losing streams, and groundwater divides may move position. The recharge variation affects the groundwater flow; for example, with a decrease of recharge, the area feeding the spring shrinks, while the area feeding the lower regional flow system increases (Bernstein et al., 2008). This implies that any decrease in annual recharge will produce a larger decrease in the spring yield and to a smaller percent decrease of the regional flow. Population growth also poses threat to recharge and that is higher than climate change. The effect of climate change on groundwater in different regions of the world is available in the literature. For example, in Swan Coastal Plain of Western Australia (Sharma, 1989); Island of Samsoe Denmark (Thomsen, 1990); northwestern United States (Vaccaro, 1992); the United Kingdom (Cooper, Wilkinson, & Arnell, 1995; Limbrick, Whitehead, Butterfield, & Reynard, 2000; Wilkinson & Cooper, 1993); Yucca Mountain (Gureghian, De Wispelare, & Sagar, 1994); Ogallala Aquifer in the central United States (Rosenberg et al., 1999); karst aquifer in South-Central Texas, United States (Rosenberg et al., 1999); France (Bouraoui, Vachaud, Li, LeTreut, & Chen, 1999); northeastern United States (Kirshen, 2002); Lansing, Michigan (Croley & Luukkonen, 2003); Germany (Eckhardt & Ulbrich, 2003); Grand Forks Aquifer in South-Central British Columbia, Canada (Scibek & Allen, 2006; Scibek, Allen, Cannon, & Whitfield, 2006); Grand River watershed, Ontario (Jyrkama & Sykes, 2007); Grote-Nete catchment, Belgium (Woldeamlak et al., 2007); upper Scezibwa catchment, Uganda (Nyenje & Batelaan, 2009); chalky groundwater basin in Belgium (Brouyere, Carabin, & Dassargues, 2004); river basins (Hupsel, Gulp, and Noor) in Netherlands, (Haugland) Norway, and (Monachyle) Scotland (Querner, Tallaksen, Kašpárek, & van Lanen, 1997); subcatchment of the Kleine Nete, Belgium (Dams et al., 2007); Oliver region of the South Okanagan, British Columbia, Canada (Toews & Allen, 2009); Mount Makiling forest (Combalicer, Cruz, & Im, 2010); a chalky aquifer, Geer basin, Belgium (Brouyere et al., 2004); Grand River watershed in Ontario, Canada (Jyrkama & Sykes, 2007); South-Central British Columbia, Canada (Scibek & Allen, 2006); Newfoundland, Canada (Bobba, Singh, Jeffries, & Bengtsson, 1997); the Netherlands (Querner, 1997); eastern Massachusetts (Kirshen, 2002); Turkey (Yagbasan, 2016); British Columbia, Canada (Allen, Mackie, & Wei, 2004); Columbia Plateau, Washington (Vaccaro, 1992); Illinois, the United States (Changnon, Huff, & Hsu,

1988); and at global level (Zektser & Loaiciga, 1993). The results of these studies portray the following:

- Effect of P_{Cp} on groundwater is positive as it increases recharge due to increased hydraulic conductivity and gradient, which increases groundwater resulting in shifts in water levels and storage.
- With increased P_{cp}, the contribution of base flow to river runoff and direct groundwater discharge to oceans increased.
- Increased P_{cp} increases soil moisture and subsequently ET, which may or may not change the recharge
- The higher intensity and frequency of P_{cp} contribute significantly to surface runoff.
- Increased temperature increases ET rates and reduces recharge. In winter, increased temperatures reduce the extent of ground frost and shift the snow melt from spring toward winter, allowing more water to infiltrate into the ground, resulting in increased groundwater recharge.
- Increased CO_2 affect the groundwater recharge by decreasing ET, increasing leaf size and crop duration. The resultant ET by the interaction of changed weather parameters and composition of the atmosphere relative to P_{cp} determine the recharge. For example, if increases in ET (due to increases in temperature) exceed increased P_{cp}, there is possibility that increases in ET could consume enough soil water to severely reduce groundwater recharge.
- Groundwater recharge is also modified by the time slice and climate change scenario. The variability in recharge to the aquifer would be more in historic data than GCM. In the future, under all scenarios of anticipated changes in temperature and CO_2 concentrations, recharge would be reduced, but reduction would be more with A1 scenario than B2.
- Seasonal change would influence groundwater; for example, increased P_{cp} during winter in Germany increases winter recharge. The effect of such recharge is counteracted during summer by the reduced recharge caused by longer-lasting soil moisture deficits.
- A thick saturated zone found at the basin smoothes the impact of seasonal variation.
- As climate changes spatially, its effect on groundwater will also vary spatially. Spatial distribution of variation in recharge of groundwater levels can be much greater than that of temporal.
- In permafrost regions, where recharge is ignored at present in global analyses may be particularly enhanced in future while coupling between surface-water and groundwater systems.

- Projected climate change shows variable groundwater recharge in different regions of the same country. For example, in the United States, the total recharge will decline across the southern aquifers, no change in the northern set of aquifers. In the mountain system, recharge is expected to decline across much of the region due to decreased snowpack.
- During the past century, changing global groundwater discharge has contributed to sea-level rise. The contribution of discharged water to sea-level rise would have been more in case of reduced water storage in land-surface reservoirs or channeled into aquifers by irrigation return flow.
- On some groundwater resources such as Edward's aquifer in Texas, United States, and Chalk aquifer in eastern England, climate change showed its effect even at normal or not increased pumping rates.
- Climate change affects groundwater discharge directly or indirectly through soil degradation, changes in water demand, and changes in irrigation or land use practices. However, the effect of land use is less at present due to irrigation.
- Though climate change affects recharge, yet, its effect is less than that of land use. Recharge may be more with vegetation than P_{cp}.

4.5 GROUNDWATER PUMPING AND GREENHOUSE GAS EMISSIONS

In arid and semiarid areas where groundwater is the only source of water, water is extracted for different purposes such as agricultural, industrial, and urban uses. The prominent challenges are groundwater reliance and decline because of overexploitation. These challenges stimulate the demand for more energy to pump water from the aquifer especially from deep groundwater resources. The energy from electricity and diesel consumption intensifies the production of large amounts of greenhouse gases especially CO_2 (Nazari, Ebadi, & Khaleghi, 2015). It is estimated that the energy required to lift $1\,m^3$ of water (with a density $= 1\,Mg\,m^{-3}$) through $1\,m$ (at 100% efficiency) is equal to $0.0027\,kWh$ (Eq. 4.4):

$$E = \frac{mgh}{Time_{cf} \times Pumping_{efficiency}} \qquad (4.4)$$

where E is energy required in kWh, m is mass of water in kg, g is acceleration due to gravity in Newton ($9.8\,ms^{-2}$), $Time_{cf}$ is the time conversion factor (3.6×10^2), and h is height of water lifting in m and can also be taken as the total dynamic head, which is the function of initial water level, drawdown, delivery head, and losses in pipe and is calculated as Eq. (4.5):

$$h = \text{initial water level} + \text{fricitonal loss} + \text{drawdown} \qquad (4.5)$$

Pumping efficiencies are around 30% (Shah, 2009). Transmission and distribution of electricity can be taken in the range of 10%–19%. The emissions of GHGs resulting from consumption of electricity are calculated according to Eq. (4.6):

$$\text{GHG emissions} = E \times \lambda_1 \qquad (4.6)$$

where GHG emissions are the amount of CO_2 emitted in kg, E is the amount of electricity used for lifting per unit volume of water (kWh m^{-3}), and λ_1 is conversion coefficients for GHG emissions per kWh of electricity consumption that is adopted to be $1.69 \, \text{kg} \, CO_2 \, \text{kWh}^{-1}$ (Nazari et al., 2015). Under traditional irrigation methods in the Punjab state of India, each meter fall in groundwater levels contributes $19 \, \text{kg carbon ha}^{-1}$ emissions through pumping (Kaur, Aggarwal, & Lal, 2016).

4.6 MODELING GROUNDWATER RESPONSE TO CLIMATE CHANGE

For rational planning of water resource systems, it is a prerequisite that changes in water requirement in response to climate change has to be predicted one or more decades ahead into the future. Coupled hydroclimate models are of help in such planning. Actually, modeling groundwater response to climate change is a very complex process because climate parameters (dominantly P_{cp}, CO_2, T_{max}, and T_{min}) affect almost all the physical, chemical, and biological processes in soil and plant, root-zone water balance, and ultimately groundwater. In recent years with the rapid increase in computation power, the wide availability of computers and model software, and the advancement in modeling, a number of models have been developed for an individual compartment, namely, climate, soil-water, crop, and groundwater.

The methodology to model groundwater consists of three main steps. In the first step, climate scenarios are formulated for the future years such as 2050 and 2100. It can be done by assigning percentage or value changes of climatic variables on a seasonal and/or annual basis only for the future years relative to the present year. In the second step, based on these scenarios and present situation, seasonal and annual recharge, evapotranspiration, and runoff are simulated with the WHI UnSat Suite (HELP module for recharge), WetSpass model, etc. In the third step, the annual recharge outputs from WHI UnSat Suite or WetSpass model are used to simulate groundwater system conditions using steady-state MODFLOW model setups for the existing conditions and for the future years. Transient simulations are undertaken to investigate the temporal response of the aquifer system to historic and future climate periods (Toews & Allen, 2009). A typical flow chart for various aspects of such a study (Fig. 4.5) shows the connections from the climate analysis to recharge simulation and, finally, to a groundwater model.

More recent techniques propose using integrated hydrologic models that can simulate the supply and demand components of all the water all the time everywhere, such as MODFLOW with the Farm Process (Schmid & Hanson, 2009). The extension of MODFLOW onto the landscape with the Farm Process (MF-FMP) facilitates fully coupled simulation of the movement of water from P_{cp}, stream flow and runoff, groundwater flow, and consumption by natural and agricultural vegetation throughout the hydrologic system at all times. The MF-FMP has the ability to simulate the conjunctive use of surface and subsurface water resources to meet agricultural and urban water demands in the presence of multiple constraints on surface water and groundwater supply (Hanson, Schmid, Faunt, & Lockwood, 2010).

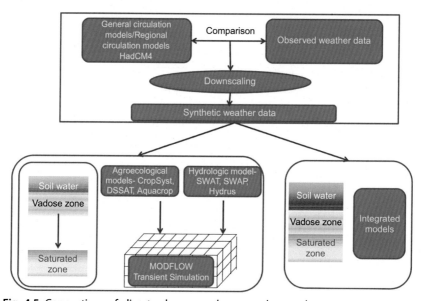

Fig. 4.5 Connections of climate change, recharge, and groundwater.

The use of integrated hydrologic model (MOHISE), consisting of soil, surface water, and groundwater submodels linked through flux boundaries, has been made by Brouyere et al. (2004) to study the impact of climate change on a chalky groundwater basin in Belgium. Many of the studies also took into consideration future land use changes using CLUE-S model to assess the impacts of climate change (Dams et al., 2007).

In a nutshell, no widely accepted methodology for groundwater modeling has emerged due to vast differences between study sites and approaches, and results vary significantly from one study to another. But to be aware of the impact of climate change on groundwater, climate, soil-water-crop, and groundwater models need to be linked. To make the methodology clear a bit, a case study of Ludhiana District in Indian Punjab is presented in which Kaur et al. (2014) and Kaur, Jalota, et al. (2015) linked atmosphere, soil-water-crop, and groundwater compartments via RCM-PRECIS, CropSyst, and MODFLOW models, respectively (Fig. 4.6), and predicted successfully water-balance components for the past years and projected groundwater storage and level and energy for pumping groundwater for future. The steps followed to project groundwater recharge, draft, and water levels as influenced by climate change are as follows.

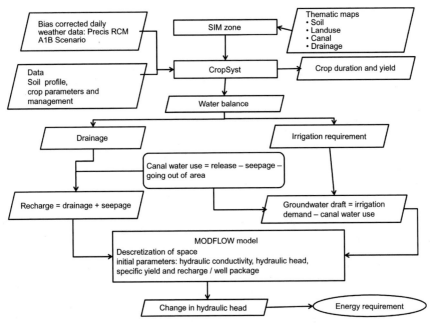

Fig. 4.6 Stepwise methodology of groundwater modeling.

4.6.1 Collection of Weather Data

For any of the study area, location/district/region, daily data on P_{cp}, T_{max}, and T_{min} recorded at different meteorological observatories for the past 30 years and under a known scenario (e.g., Special Report on Emission Scenarios (SRES) for baseline (1961–90), midcentury (2021–50), and end century (2071–98)) derived from coupled global climate/regional climate model have to be collected. Although forecasts of future climate events by the models are still imprecise, they nevertheless may represent the best available decision-making information at the present time. If any climate parameter is missing, that is generated by weather generator, for example, solar radiation with Hargreaves and Samani (1982), relative humidity and wind speed by ClimGen (Stockle & Nelson, 1999), and WGEN (Richardson & Wright, 1984) models. The magnitude and trends of the climate modeled data may not match the observation. Such discrepancy can be minimized by statistical bias correction methods (discussed in Chapter 2) for its use in weather file of the soil-water-crop model. Data for CO_2 levels for a given year in future and scenario can be taken from Bern climate change model (http://www.ipcc-data.org/observ/ddc_co2.html) or other models.

4.6.2 Development of Unique Simulation Units

To account for soil types, land use, and surface-water supply heterogeneity in the area, various thematic maps such as soil, land use, canal network, and drainage maps are digitized in GIS software to create independent layers. The maps available with line shapefile data such as of canal and drainage are converted to a polygon map in accordance to their size. A combined map of soil type, land use, and canal and drainage network is obtained by performing various overlay operations in GIS to form a unique simulation zone. Each simulation zone is considered homogeneous with respect to soil type, land use, and drainage. However, the overlay operation may led to a large number of spurious small polygons that have no real meaning, those have to be merged with the adjoining larger units (Burrough & Rachael, 1998). This process will result in generation of unique simulation units for the entire study area. For each unique simulation unit, independent root-zone water-balance components are obtained from soil-water-crop model (CropSyst, DSSAT, APSIM, etc.), and for the entire simulation, zone is estimated by joining to polygon attribute table of GIS layer.

4.6.3 Simulation of Root Zone Water Balance

Simulations for estimation of water balance in root zone can be run using any soil-water-crop model, but that should be properly calibrated and validated and have minimum simulation error. Using weather file from corrected weather data and atmospheric CO_2 concentration for a given year and scenario from models, soil files consisting of soil physical and chemical properties, management file consisting of date of sowing, time and amount of irrigation, and nitrogen fertilizer applied, root-zone water-balance components are simulated for a given cropping system. In the model, changed CO_2 affects growth ratio, a ratio of potential growth at specified CO_2 concentration to base level by altering daily crop radiation use efficiency (Stockle, Williams, Rosenberg, & Jones, 1992; Tubiello, Donatelli, Rosenzweig, & Stockle, 2000). From the cropped area, crop duration, irrigation requirement, ET, and soil water evaporation for the cropped and fallow land are worked out (Jalota, Kaur, Kaur, & Vashisht, 2013). Similarly, the most suitable method for other land uses (Kaur et al., 2014) and runoff (curve number or the method of Batelaan & Woldeamlak, 2007) is used. Estimates of the root-zone water-balance are made from Eq. (3.11) given in Chapter 3. The drainage passing root zone is considered equivalent to the potential groundwater recharge.

4.6.4 Groundwater Recharge and Draft

The gross groundwater recharge for each simulation zone is estimated as the sum of the waters (i) that moved beyond the root zone as obtained from soil-water-crop model; (ii) seepage losses from canal, river, and drains in proportionate to the length of canal passing through the areas as per the norms provided by GWREC (1997); and (iii) change in groundwater storage due to influx and out flux from the adjoining aquifers. Spatially, varying groundwater draft from the cropped land can be obtained by subtracting canal water supplies and from built-up land (civic water use) based on population and per capita per day requirement.

4.6.5 Simulation of Groundwater

The groundwater is generally simulated with MODFLOW model (McDonald & Harbaugh, 1996) under a PMWIN environment, which is based on the three dimensional movement of groundwater in a heterogeneous and anisotropic aquifer and is given by the partial differential equation (Eq. 4.7):

$$\frac{\partial}{\partial x}\left(K_{xx}\frac{\partial h}{\partial x}\right) + \frac{\partial}{\partial y}\left(K_{yy}\frac{\partial h}{\partial y}\right) + \frac{\partial}{\partial z}\left(K_{zz}\frac{\partial h}{\partial z}\right) \pm w = S_s\frac{\partial h}{\partial t} \qquad (4.7)$$

where K_{xx}, K_{yy}, and K_{zz} are the hydraulic conductivity along the x, y, and z coordinate axes, $m\,day^{-1}$; h is the hydraulic head, m; w is the volumetric flux per unit volume and represents sources and/or sinks of water, day^{-1}; S_s = specific storage of the porous material, m^{-1}; and t is time, day.

To run MODFLOW model, firstly, the study area is discretized with constant grid spacing, generally of $2.5 \times 2.5\,km$. Then, the area is identified to know which side is bounded by a river or other water reservoirs. For that side, a constant head boundary is applied, in which cells are of constant head. The head is specified in advance and is held constant at a specified value throughout all time steps of the simulation. For constant head cells, the storage terms are not used, but the other parameters of the Eq. (4.7) are considered. For the sides that have not constant head, flux boundaries are simulated by use of appropriate recharge/well package. Darcy's law is used to compute the flux from/to individual boundary cell with the help of Arc Hydro toolkit GIS and Microsoft Excel. The cells lying inside the study area are simulated as active cells for which hydraulic head are computed throughout all time steps of the simulation. The spatial hydraulic parameters used are initial hydraulic head, hydraulic conductivity, and specific yield. For each time step after the first, the hydraulic head distribution at the start of one time step is set equal to the hydraulic head distribution at the end of previous time step. The information regarding hydraulic parameters such as hydraulic conductivity and specific yield is used either observed or estimated indirectly with the help of well logs scattered all across the study area. These point values along with the bottom elevation are interpolated in GIS to obtain spatial distribution of each parameter. The temporal parameters, that is, recharge and drafts, are simulated using well and recharge flow packages for each of the discretized cell. The interpolated rasters are converted to American Standard Code for Information Interchange (ASCII) files to enable easy import the values for the tables in PMWIN. The headers created in the ASCII file by ArcMap are not compatible with PMWIN, and they can be manually modified.

However, to know the effects of climate variability and change on future water resources in more precise way, advanced research needs the following considerations (Koch, 2008):

- Knowing the response of plant transpiration to increased CO_2, climate warming and changes in soil moisture, and its inclusion in recharge models for groundwater changes

- Evaluating and improving global climate models in terms of the most critical parameters for hydrology, such as extremes of precipitation, and ET incorporating humidity, cloudiness, and radiation
- Representing yearly to decadal-scale climate variability in global climate models through representations of driving mechanisms such as El Niño, the Interdecadal Pacific Oscillation for the Pacific region and the Northern Atlantic Oscillation (NAO) and the Arctic Oscillation for the northern Atlantic hemisphere
- Reducing uncertainty in climate projections through further research for particular applications such as hydrology
- Separating anthropogenic-induced changes from natural climate change and variability by including high-resolution paleorecords of hydrologic and climatological parameters
- Developing and applying downscaling methods that represent climate at the relatively fine spatial and temporal scales of landscape hydrology
- Representing the continental physical hydrology in global climate models to simulate the interactions between climate and hydrology
- Understanding the large-scale physical hydrology and its effects on the subsurface recharge process to be able to project future hydrologic behavior for unprecedented climate conditions
- Improving the understanding of the interactions of groundwater with land and surface-water resources by developing better integrated surface-water/groundwater models
- Getting a better hold on the importance of feedbacks through vegetation on hydrology and how these might change under future climate and CO_2 concentrations
- Increasing use of remotely sensed data for climatological and hydrologic applications as from the very promising GRACE earth satellite project
- Generating more information on groundwater recharge mechanisms, storage capacity, and residence times in extremely cold and alpine conditions

4.7 GROUNDWATER MANAGEMENT TO MINIMIZE CLIMATE CHANGE IMPACT

Proper management of groundwater is vital to (i) minimize the impact of climatic variation and the worst drought situations caused by climate change and (ii) protect groundwater resources as the aquifers having a high storage capacity are less sensitive to climate change than surface-water bodies. As the response of groundwater to climate change is slower than surface-water system, therefore, management of groundwater systems needs ahead planning. The groundwater can be managed by minimizing the gap between demand and supply, that is, limiting the expansion of water supply from the aquifer and increasing the groundwater recharge, exploiting feasible deep aquifers, and reducing groundwater discharging from store as surface-water base flow. Specific measures to minimize the effect of climate change on groundwater resources are as follows:

- Use the simple and low-cost traditional techniques like water harvesting, preservation or restoration of natural infiltration conditions, and increase artificial groundwater recharge through drainage channel and recharge ponds. The success of augmentation of groundwater resources through artificial recharge depends upon the availability of good-quality water, suitability of the site, and recharge technique. The water (of good quality) that can be used for recharge includes surplus P_{cp}, canal water during rainy season, and treated sewage water. In general, artificial recharge is recommended under the conditions of (i) unconfined permeable and sufficiently thick aquifer; (ii) sufficiently permeable surface soil to maintain high infiltration rate; (iii) gently sloping land; (iv) sufficiently deep groundwater level to accommodate water-table rise; and (v) moderate hydraulic conductivity of aquifer material. The methods used for artificial recharge are direct methods like flooding, ditch and furrows, recharge basins, runoff conservation structures, injection wells, recharge wells, surface irrigation and by indirect methods like pumping wells, collector wells, bore blasting, and hydrofracturing. The choice of recharging technique is site-specific (GOI, 2007) and is given in Table 4.1.
- Manage recharge by LULC changes keeping in view that vegetation acts in both ways, that is, increasing and decreasing recharge.
- Afforest the area depending upon the climate zone tropical arid and semiarid (Bala et al., 2007).
- Increase rainwater harvesting by catching runoff in basins or by infiltration wells.
- Divert canal water to un–utilized or choked drains after repair.
- Modify village ponds.
- Construct check structures across the drains and divert flood water into these.
- Increase height of bunds around the paddy fields.
- Develop an early warning system that monitors changing climatic conditions and triggers contingency plans for avoiding crop failure.

Table 4.1 Methods of Artificial Recharge in Relation to Site Characteristics

Method	Requirements/Site Characteristics
1. Surface spreading technique	
1.1 Flooding — Spreading of surplus surface water from canals over large area for a sufficient period	Availability of sufficiently large area adjacent to the canal/stream
1.2 Ditches and furrow	Areas with irregular topography
1.3 Recharge basins — Either excavated or enclosed by dikes or levees — Commonly built parallel to streams/canals	(a) Periodic maintenance such as scraping (b) Water released to basin should have minimum sediment
1.3 Percolation tanks — Small tanks created by making low elevation stop dams across streams or located adjacent to a stream by excavation and connected to the stream by delivery canal	(a) High permeability of rocks coming under submergence areas (b) Uniform degree and extent of weathering of rocks
1.5 Streams/channels modification	Influent stream bed above water table
2. Subsurface techniques	Deeper aquifers overlain by impervious layers, the infiltration from surface cannot recharge the subsurface aquifer under natural conditions
2.1 Injection wells — Treated surface water is pumped in	Adequately filtered and disinfected water for recharge
2.2 Gravity head recharge wells — Ordinary bore wells, tube wells, and dug wells used for pumping	–do–
2.3 Connector wells — Special type of wells by which water is made flow from one aquifer to other without any pumping due to difference in piezometric head	Multiple aquifer system, deeper aquifer having lower piezometric head than lower lying aquifers separated by impermeable confining layer
2.4 Recharge pits	
2.5 Recharge shafts similar to a recharge pit but smaller in diameter	Impermeable layers or lenses form a barrier between the surface water and the water table. Water-table aquifer located deep below land surface is overlain by poorly permeable strata

Table 4.1 Methods of Artificial Recharge in Relation to Site Characteristics—cont'd

Method	Requirements/Site Characteristics
3. Indirect methods	
3.1 Induced recharge – Pumping wells, tube wells, dug wells	Aquifer hydraulically connected with surface water
3.2 Ground water dam/subsurface dikes	(a) Stream is in hydraulic connection with the phreatic aquifer (b) Valley should be well defined and wide with narrow outlet (bottom necked) (c) Occurrence of unconfined aquifers within a shallow to moderate depth (10–20 m below ground level) (d) Adequate thickness of aquifer underlying the site (minimum of 5 m)
4. Combination methods	
4.1 Recharge basin with shaft	Rocks exposed on the bottom of basin are not permeable enough to allow the stored water to infiltrate at a fast rate
4.2 Induced recharge with connector wells	Area comprising of multiple aquifer system existing adjacent to perennial river. Connector well is used to connect two aquifers

- Examine the climate change impact on groundwater resources by indirect effects like changes in land use, irrigation, and groundwater exploitation and optimize those with the modeling approach.

The other measures for groundwater management are enhancing the surface-water supply and reducing groundwater draft. The surface-water supply can be enhanced by (i) developing new hydroprojects; (ii) interzonal transfer of surplus water; (iii) renovation of waste water; and (iv) increasing canal water supply to declining water-table areas, which can be helpful in recharging the groundwater and meeting partial demand of irrigation. Agricultural sector being the main user of groundwater water requires more attention. Groundwater draft in agriculture can be reduced by the following:

- Applying less irrigation water and increasing water application efficiency with methods such as land leveling by laser.
- Optimizing plot size and use of microirrigation techniques, better scheduling and method of irrigation, and use of straw/crop residue as mulch (Package of Practices, PAU; http://web.pau.edu/index.php?_act=manageLink&DO=firstLink&intSubID=99). These methods will reduce not only the withdrawal of groundwater but also use of energy. Cultivation of drought-resistant plants also reduces groundwater draft.
- Increasing surface-water supply and its entry into the soil from canals, reducing runoff, and increasing infiltration of the surface soil by modifying soil surface conditions and drainage of P_{cp} water in vadose zone.
- Encouraging real water saving by reducing ET loss particularly soil water evaporation through synchronizing crop period with lower evaporative demand, adopting shorter duration hybrid crops, using mulching in summer and spring crops, replacing high ET with low ET crops (crop diversification), and reducing unproductive water loss by soil water evaporation by shallow tillage and straw mulching (Jalota et al., 2013, 2009, 2011; Jalota & Arora, 2002; Jalota & Prihar, 1998).
- Increasing groundwater recharge in hilly areas by proper gully erosion management. Managing first-order gullies in a catchment helps in increasing residence time for runoff water on soil surface, which enhances its infiltration into the soil, and recharging the groundwater on a long-term basis.

EXERCISES

1. Calculate actual electricity used to lift $1000\,m^3$ through height of $1\,m$ if electric irrigation pumps operate at 40% efficiency and transmission and distribution losses in delivering power to pump sets are of the order of 25%. Also calculate carbon dioxide emission.

   Answer $(2.7\,kWh^{-1}$ and $4.6\,kg\,CO_2)$.
2. Discuss how climate change, anthropogenic activities, and aquifer properties affect the recharge process.
3. Is vegetation beneficial for recharge, support your answer with processes involved?
4. What are the categories of groundwater resources/aquifers in relation to their sensitivity to climate change effect?
5. Discuss the various techniques of recharge measurement in relation to hydraulic sources from which data is taken.

6. What are different ways by which sustainability of groundwater is affected by climate change?

7. What type of advance research is needed to understand the effects of climate variability and change on future water resources availability?

8. Under which conditions artificial recharge is recommended for managing groundwater? Is artificial recharge site-specific, discuss?

9. Discuss the ways by which draft can be reduced in agriculture sector.

10. Calculate the change on water level from the data on ET and P_{cp} in different crops, assuming that 75% of the deficit is met from groundwater and specific yield is 0.20.

	Crop					
	Rice	Soybean	Maize	Cotton	Wheat	Chickpea
ET (mm)	641	697	485	802	405	326
P_{cp} (mm)	566	626	572	649	115	125

Answer ($-0.28, -0.27, 0.33, -0.57, -1.09$, and -0.75 m in rice, soybean, maize, cotton, wheat, and chickpea, respectively).

REFERENCES

Acreman, M. (2000). *Hydrology of the UK*: (p. 336). United Kingdom: Routledge.

Akhtar, M., Ahmed, N., & Booij, M. J. (2008). Impact of climate change on water resources of Hindukush-Karakorum-Himalaya region under different glacier coverage scenarios. *Journal of Hydrology, 355*, 148–163.

Allen, D. M., Mackie, D. C., & Wei, M. (2004). Groundwater and climate change: A sensitivity analysis for the Grand Forks aquifer, southern British Columbia, Canada. *Hydrogeology Journal, 12*, 70–290.

Anuraga, T. S., Ruiz, L., Kumar, M. S., Sekhar, M., & Leijnse, A. (2006). Estimating groundwater recharge using land use and soil data: A case study in South India. *Agricultural Water Management, 84*, 65–76.

Bala, G., Caldeira, K., Wickett, M., Phillips, T. J., Lobell, D. B., Delire, C., et al. (2007). Combined climate and carbon-cycle effects of large-scale deforestation. *PNAS (physical sciences/environmental sciences), 104*, 6550–6555.

Batelaan, O., De Smedt, F., & Triest, L. (2003). Regional groundwater discharge: Phreatophyte mapping, groundwater modelling and impact analysis of land-use change. *Journal of Hydrology, 275*, 86–108.

Batelaan, O., & Woldeamlak, S. T. (2007). *A review interface for WetSpass*: (p. 50). Brussel: Department of hydrology and hydraulic engineering, Vrije Universiteit Brussel.

Bellot, J., Bonet, A., Sanchez, J. R., & Chirino, E. (2001). Likely effects of land use changes on the runoff and aquifer recharge in a semiarid landscape using a hydrological model. *Landscape and Urban Planning, 55*, 41–53.

Bernstein, L., Bosch, P., Canziani, O., Chen, Z., Christ, R., & Riahi, K. (2008). *Climate change 2007: Synthesis report*. Intergovernmental Panel on Climate Change.

Bobba, A. G., Singh, V. P., Jeffries, D. S., & Bengtsson, L. (1997). Application of a watershed runoff model to north-east Pond River, Newfoundland: To study water balance and hydrological characteristics. *Hydrological Processes*, *11*, 1573–1593.

Bouraoui, F., Vachaud, G., Li, L. Z. X., LeTreut, H., & Chen, T. (1999). Evaluation of the impact of climate changes on water storage and groundwater recharge at the watershed scale. *Climate Dynamics*, *15*, 153–161.

Brouyere, S., Carabin, G., & Dassargues, A. (2004). Climate change impacts on groundwater resources: Modelled deficits in a chalky aquifer, Geer basin, Belgium. *Hydrogeology Journal*, *12*, 123–134.

Burnett, W. C., Aggarwal, P. K., Aureli, A., Bokuniewicz, H., Cable, J. E., Charette, M. A., et al. (2006). Quantifying submarine groundwater discharge in the coastal zone via multiple methods. *Science of the Total Environment*, *367*, 498–543.

Burrough, P. A., & Rachael, A. M. (1998). *Principles of geographical information systems* [pp. 333]. Oxford: Oxford University Press.

Changnon, S. A., Huff, F. A., & Hsu, C. F. (1988). Relations between precipitation and shallow groundwater in Illinois. *Journal of Climate*, *1*, 1239–1250.

Chen, J., Li, J., Zhang, Z., & Ni, S. (2014). Long-term groundwater variation in Northwest India from Satellite. *Global and Planetary Change*, *116*, 130–138.

Combalicer, E. A., Cruz, R. V. O., & Im, S. (2010). Assessing climate change impacts on water balance in the Mount Makiling forest. *Journal of Earth System Science*, *119*, 265–283.

Cooper, D. M., Wilkinson, W. B., & Arnell, N. W. (1995). The effects of climate changes on aquifer storage and river baseflow. *Hydrological Sciences Journal*, *40*, 615–632.

Croley, T. E., & Luukkonen, C. L. (2003). Potential effects of climate change on ground water in Lansing, Michigan. *Journal of the American Water Resources Association*, *39*, 149–163.

Dams, J., Woldeamlak, S. T., & Batelaan, O. (2007). Forecasting land-use change and its impact on the groundwater system of the Kleine Nete catchment, Belgium. *Hydrology and Earth System Sciences Discussions*, *4*, 4265–4295.

Doll, P., Dobreva, H. H., Portmanna, F. T., Siebert, S., Eicker, A. A., Rodell, M. C., et al. (2012). Impact of water withdrawals from groundwater and surface water on continental water storage variations. *Journal of Geodynamics*, *59–60*, 143–156.

Dragoni, W., & Sukhija, B. S. (2008). Climate change and groundwater—A short review. In W. Dragoni & B. S. Sikhija (Eds.), *Climate change and groundwater* (pp. 1–12). London: Geological Society. Special Publications 288.

Eckhardt, K., & Ulbrich, U. (2003). Potential impacts of climate change on groundwater recharge and stream flow in a central European low mountain range. *Journal of Hydrology*, *284*, 244–252.

Finch, J. W. (2001). Estimating change in direct groundwater recharge using a spatially distributed soil water balance model. *Quarterly Journal of Engineering Geology and Hydrogeology*, *34*, 71–83.

GOI (2007). *Manual on artificial recharge of ground water*. India: Central Ground Water Board, Ministry of Water Resources.

Green, T. R., Taniguchi, M., Kooi, H., Gurdak, J. J., Allen, D. M., Hiscock, K. M., et al. (2011). Beneath the surface of global change: Impacts of climate change on groundwater. *Journal of Hydrology*, *405*, 532–560.

Gureghian, A. B., De Wispelare, A. R., & Sagar, B. (1994). In *Vol. 3. Sensitivity and probabilistic analyses of the impact of climatic conditions on the infiltration rate in a variably saturated multilayered geologic medium Proceedings of the fifth international conference on high level radioactive waste management* (pp. 1622–1633): American Nuclear Society.

GWREC (1997). *Groundwater resource estimation methodology—(1997)*. Report of the groundwater estimation committee New Delhi: Ministry of water resources, Government of India.

Hanson, R. T., Flint, L. E., Flint, A. L., Dettinger, M. D., Faunt, C. C., Cayan, D., et al. (2012). A method for physically based model analysis of conjunctive use in response to potential climate changes. *Water Resources Research, 48,* 1–23.

Hanson, R. T., Newhouse, M. W., & Dettinger, M. D. (2004). A methodology to asess relations between climatic variability and variations in hydrologic time series in the southwestern United States. *Journal of Hydrology, 287,* 252–269.

Hanson, R. T., Schmid, W., Faunt, C. C., & Lockwood, B. (2010). Simulation and analysis of conjunctive use with MODFLOW's farm process. *Ground Water, 48,* 674–689.

Hargreaves, G. H., & Samani, Z. A. (1982). Estimating potential evapotranspiration. *Journal of Irrigation and Drainage Engineering, 108,* 223–230.

Jalota, S. K., & Arora, V. K. (2002). Model based assessment of water balance components under different cropping systems in north-west India. *Agricultural Water Management, 57,* 75–87.

Jalota, S. K., Kaur, H., Kaur, S., & Vashisht, B. B. (2013). Impact of climate change scenario on yield, water and nitrogen—balance and—use efficiency of rice-wheat cropping system. *Agricultural Water Management, 116,* 29–38.

Jalota, S. K., & Prihar, S. S. (1998). *Reducing soil water evaporation by tillage and straw mulching.* Ames: IOWA State University Press.

Jalota, S. K., Singh, K. B., Chahal, G. B. S., Gupta, R. K., Chakraborty, S., Sood, A., et al. (2009). Integrated effect of transplanting date, cultivar and irrigation on yield, water saving and water productivity of rice (*Oryza sativa* L.) in Indian Punjab: Field and simulation study. *Agricultural Water Management, 96,* 1096–1104.

Jalota, S. K., Vashisht, B. B., Kaur, H., Arora, V. K., Vashist, K. K., & Deol, K. S. (2011). Water and Nitrogen—balance and—use efficiency in a rice (Oryza sativa)-wheat (Triticum aestivum) cropping system as influenced by management interventions: Field and simulation study. *Experimental Agriculture, 47,* 609–628.

Jyrkama, M. I., & Sykes, J. F. (2007). The impact of climate change on spatially varying groundwater recharge in the grand river watershed. *Journal of Hydrology, 338,* 237–250.

Kaur, S., Aggarwal, R., Jalota, S. K., & Sood, A. (2015). Estimation of fluxes across boundaries for groundwater flow model using GIS. *Current Science, 109,* 607–610.

Kaur, S., Aggarwal, R., Jalota, S. K., Vashisht, B. B., & Lubana, P. P. S. (2014). Estimation of groundwater balance using soil-water-vegetation model and GIS. *Water Resources Management, 28,* 4359–4371.

Kaur, S., Aggarwal, R., & Lal, R. (2016). Assessment and mitigation of greenhouse gas emissions from groundwater irrigation. *Irrigation and Drainage, 65,* 762–770.

Kaur, S., Jalota, S. K., Singh, K. G., Lubana, P. P. S., & Aggarwal, R. (2015). Assessing climate change impact on root-zone water balance and groundwater levels. *Journal of Water and Climate Change, 6,* 436–448.

Kirshen, P. H. (2002). Potential impacts of global warming on groundwater in eastern Massachusetts. *Journal of Water Resources Planning and Management, 128,* 216–226.

Koch, M. (2008). *Challenges for future sustainable water resources management in the face of climate change.* http://www.uni-kassel.de/fb14/geohydraulik/Koch/paper/2008/Nakon_Pathom/Climate_Change_Groundwater_Abs.pdf.

Kundzewicz, Z. W., & Doell, P. (2009). Will groundwater ease freshwater stress under climate change? *Hydrological Sciences Journal, 54,* 665–675.

Lerner, D. N. (2002). Identifying and quantifying urban recharge: A review. *Hydrogeology Journal, 10,* 143–152.

Lerner, D., Issar, A. S., & Simmers, I. (Eds.), (1990). *Groundwater recharge: A guide to understanding and estimating natural recharge* (p. 345). Hannover, FRG: Heise.

Limbrick, K. J., Whitehead, P. G., Butterfield, D., & Reynard, N. (2000). Assessing the potential impacts of various climate change scenarios on the hydrological regime of the River Kennet at Theale, Berkshire, south-centra England, UK: An application and evaluation of the new semi-distributed model, INCA. *Science of the Total Environment*, *251/252*, 539–556.

McDonald, M. G., & Harbaugh, A. W. (1996). *A modular three dimensional finite difference groundwater flow model*. United States Geological Survey.

Nazari, S., Ebadi, T., & Khaleghi, T. (2015). Assessment of nexus between groundwater extraction and green house gas emissions employing aquifer modelling. *Procedia Environmental Sciences*, *25*, 183–190.

Nyenje, P. M., & Batelaan, O. (2009). Estimating the effects of climate change on groundwater recharge and base flow in the upper Ssezibwa catchment, Uganda. *Hydrological Sciences Journal*, *54*, 713–726.

Querner, E. P. (1997). Description and application of the combined surface and groundwater flow model MOGROW. *Journal of Hydrology*, *192*, 158–188.

Querner, E. P., Tallaksen, L. M., Kašpárek, L., & van Lanen, H. A. J. (1997). Impact of land-use, climate change and groundwater abstraction on stream flow droughts basin using physically-based models. In A. Gustard, S. Blaskova, M. Brilly, S. Demuth, J. Dixon, H. van van Lanen, C. Llasat, S. Mkhandi, & E. Servat (Eds.), *FRIEND'97—Regional hydrology: Concepts and models for sustainable water resource management* (pp. 171–179). IAHS Publication no. 246.

Richardson, C. W., & Wright, D. A. (1984). *WGEN: A model for generating daily weather variables*. USDA, ARS–8, pp. 83.

Rodell, M., Velicogna, I., & Famiglietti, J. S. (2009). Satellite-based estimates of groundwater depletion in India. *Nature*, *460*(7258), 999–1002.

Rosenberg, N. J., Epstein, D. J., Wang, D., Vail, L., Srinivasan, R., & Arnold, J. G. (1999). Possible impacts of global warming on the hydrology of the Ogallala Aquifer Region. *Climate Change*, *42*, 677–692.

Scanlon, B. R., Healy, R. W., & Cook, P. G. (2002). Choosing appropriate techniques for quantifying groundwater recharge. *Hydrogeological Journal*, *10*, 18–39.

Scanlon, B. R., Keese, K. E., Flint, A. L., Flint, L. E., Gaye, C. B., Edmunds, W. M., et al. (2006). Global synthesis of groundwater recharge in semiarid and arid regions. *Hydrological Processes*, *20*, 3335–3370.

Schmid, W., & Hanson, R. T. (2009). *The farm process version 2 (FMP2) for MODFLOW–2005: Modifications and upgrades to FMP1*. United States Geological Survey.

Scibek, J., & Allen, D. M. (2006). Comparing modelled responses of two high-permeability, unconfined aquifers to predicted climate change. *Global and Planetary Change*, *50*, 50–62.

Scibek, J., Allen, D. M., Cannon, A. J., & Whitfield, P. (2006). Groundwater–surface water interaction under scenarios of climate change using a high-resolution transient groundwater model. *Journal of Hydrology*, *333*, 165–181.

Shah, T. (2009). Climate change and groundwater: India's opportunities for mitigation and adaption. Environmental Research Letters, *4*, 035005.

Sharif, M. M., & Singh, V. P. (1999). Effect of climate change on sea water intrusion in coastal aquifers. *Hydrological Processes*, *13*, 1277–1287.

Sharma, M. L. (1989). *Groundwater recharge*. Rotterdam: AA Balkema.

Stockle, C. O., & Nelson, R. (1999). *ClimGen. A weather generator program*. Pullman, WA: Biological system Engineering Department, Washington State University.

Stockle, C. O., Williams, J. R., Rosenberg, N. J., & Jones, C. A. (1992). A method for estimating the direct and climatic effects of rising atmospheric carbon dioxide on growth and yield of crops: Part I—Modification of the EPIC model for climate change analysis. *Agricultural Systems*, *38*, 225–238.

Taylor, R. G., Scanlon, B., Doll, P., Rodell, M., Van Beek, R., Wada, Y., et al. (2013). Ground water and climate change. *Nature Climate Change*, *3*, 322–329.

Thomsen, R. (1990). Effect of climate variability and change in groundwater in Europe. *Nordic Hydrology*, *21*, 185–194.

Toews, M. W., & Allen, D. M. (2009). Simulated response of groundwater to predicted recharge in a semi-arid region using a scenario of modelled climate change. *Environmental Research Letters*, *4*, 1–19.

Tubiello, F. N., Donatelli, M., Rosenzweig, C., & Stockle, C. O. (2000). Effects of climate change and elevated CO_2 on cropping systems: Model predictions at two Italian locations. *European Journal of Agronomy*, *13*, 179–189.

Vaccaro, J. J. (1992). Sensitivity of groundwater recharge estimates to climate variability and change, Columbia Plateau, Washington. *Journal of Geophysical Research Atmospheres*, *97*, 2821–2833.

Walker, G. R., Zhang, L., Ellis, T. W., Hatton, T. J., & Petheram, C. (2002). Towards a predictive framework for estimating recharge under different land uses: Review of modeling and other approaches. *Hydrogeology Journal*, *10*, 68–90.

Wilkinson, W. B., & Cooper, D. M. (1993). Response of idealized aquifer/river systems to climate change. *Hydrological Sciences Journal*, *38*, 379–389.

Woldeamlak, S. T., Batelaan, O., & Smedt, F. D. (2007). Effects of climate change on the groundwater system in the Grote-Nete catchment, Belgium. *Hydrogeology Journal*, *15*, 891–901.

Yagbasan, O. (2016). Impacts of climate change on groundwater recharge in Küçük Menderes River Basin in Western Turkey. *Geodinamica Acta*, *28*, 209–222.

Zektser, I. S., & Loaiciga, H. A. (1993). Groundwater fluxes in the global hydrologic cycle: Past, present, and future. *Journal of Hydrology*, *144*, 405–427.

FURTHER READING

IPCC. (n.d.). *Bern climate change model.* www.ipcc-data.org/ancillary/tar-bern.txi (Accessed 4 September 2016).

McCallum, J. L., Crosbie, R. S., Walker, G. R., & Dawes, W. R. (2010). Impacts of climate change on groundwater in Australia: A sensitivity analysis of recharge. *Hydrogeology Journal*, *8*, 1625–1638.

CHAPTER FIVE

Adaptation and Mitigation

Contents

5.1 INTRODUCTION

It is unequivocal that in the future, food production system would be under pressures of (i) ever-increasing population; (ii) changing their diets due to increased income; (iii) less land, water, and nutrients; (iv) conservation of other land uses like forests and wetlands; (v) essential ecosystem services like carbon storage and biodiversity; and (vi) climate change (warmer temperatures and misrepresented precipitation (P_{cp}) patterns). Although all these pressures are equally important, climate blues (increased temperature, drought, and floods) challenge the crop production more. Under such conditions, the current crop production systems need adaptation and mitigation to (i) overcome these pressures and (ii) reduce their carbon footprint, measured as direct emissions of gases into the atmosphere that cause climate change. Adaptation and mitigation are two policy options having the same goal; however, adaptation is increasing attention as alternative response strategy than reducing net greenhouse gas (GHG) termed as mitigation. The core challenge of climate change adaptation and mitigation in agriculture is to produce more under volatile production conditions and with net reductions in GHG emissions from food production and marketing. Some adaptations may have implications for mitigation, which are related to energy use.

In this chapter, the basics of adaptation technologies to combat the climate change impact on crop production and mitigation strategies to reduce GHG emissions are talked about. Adaptation technologies discussed are adjusting trans-planting date, creating heat tolerance in plants for reducing temperature impact, straw mulching for conserving soil moisture, and application of balanced fertilizers for making a plant healthier and stay green to tolerate volatile weather conditions. It also highlights means to reduce GHG emissions such as using solid organic N fertilizers, tightening the N cycle, minimizing fallow and bare soil emissions, and using slow-release N fertilizers for nitrous oxide (N_2O); sequestering carbon, reducing soil disturbance, practicing of organic farming, and growing cover crops for carbon dioxide (CO_2); and proper management of fertilizer (using SO_4^{2+}-containing N and compost and applying N fertilizers as foliar spray), water (intermittent irrigation water), and crop (growing upland crops and selecting rice cultivars that have high root oxidative activity) for methane (CH_4).

5.2 ADAPTATION

Adaptation is simply making adjustments to create the conditions more suitable and economical by altering or modifying both the process of adapting and the conditions being adapted. It is an essential module in assessment of climate impacts and development of adaptation policies. Adaptation is perceived from the information on (i) how much is the impact of climate change and its seriousness and (ii) how and under what conditions adaptation is expected to occur. In climate change studies, though a number of definitions of adaptation have been proposed, the common thing in them is the adjustments in a system in response to climate stimuli such as variability, change, and recurring extremes (drought, floods, storms, etc.). Nowadays, focus on adaptation is increasing because of reasons like (i) lacking in progress of global emission-reduction agreements; (ii) faster and nonlinear increase in atmospheric CO_2 and temperature and erratic P_{cp}; and (iii) their more negative impacts on crop yield and natural resources than the previous. All of these warrant understanding of the basics of adaptation and mitigation, which is the strength of this chapter.

5.2.1 Assumptions of Adaptation

Prior to selection of an adaptation measure, it is important that the managements and managers are convinced and confident that (i) projected climate changes are genuine and are likely to persist; (ii) anticipated changes will significantly impact their activity; (iii) scientific and other choices essential to counter the projected changes are available; (iv) if climate impacts necessitate major land-use change in anywhere, there may be demands to support changeovers such as relocation of industry and migration of people in that area; (v) there is a provision to develop new infrastructure, policies, and institutions to support new management and land-use arrangements; and (vi) the developed policies should be flexible, so that they can be modified in accordance to climate change by "learning by doing."

5.2.2 Classification of Adaptations

Adaptation may be classified based on the following.

Spontaneity

Autonomous: It is at an individual farmer level and involves changes in agricultural practices through his experience. It is automatic and spontaneous.

Planned: It is at a government level that resulted from the policy decision and is strategic. It may be developing infrastructures (irrigation and markets), developing technical improvement through research, and providing information to the farmer such as suitable crop and optimum time of sowing.

In anticipation: It is for the adverse consequences associated with climate change, irrespective of the individual or government level.

Time and Time Frame

Short term (tactical): Short-term climate adaptations by farmers are based on locally observed climate trends and projected climate changes from daily to interannual scales. Decisions may be tactical, which are made for short times and can be revised/adjusted frequently.

Long range (strategic): Long-term adaptations are decided based on the projections of climate and uncertainties at the finer spatial and temporal scales. It takes considerable planning and has impact over longer period (e.g., decadal timescale). Strategies are developed by knowing short-term response strategies and their relation to long-term choices to ensure that decisions implemented for the next 1–3 decades persist later in the century. Another viewpoint is that executing effective adaptation can "buy time" until an effective mitigation response is accumulated.

Response

Reactive: Reactive adaptation is when adaptation is due to immediate response to climate change. Reactive adaptation does not give the best response when our past understanding doesn't correspond to the current environmental and socioeconomic conditions.

Proactive: Proactive adaptation involves long-term decision-making that improves our ability to cope with future climate change. Periodic assessment and risk management strategies help make this response the most effective. Such adaptation is more likely to reduce long-term damage, risk, and vulnerability due to climate change.

Passive: The adaptation can be unreceptive, giving no response.

Anticipatory: The adaptation is responsive and defensive in nature.

Pathway/Progress

Incremental: When minor adjustments are made more frequently allowing for current farming objectives to be met under changed conditions, the adaptation is incremental. Such adaptation occurs if climate change remains small relative to season-to-season variation and the impact of climate is modest.

Ad hoc: Here, adjustments are made temporarily and rarely.

Effect

Here, adaptation is based on the type of effect, whether adaptation is for ameliorating the adverse consequences, reducing vulnerability, preventing the loss, tolerating the loss, etc. However, prevention of one process may affect the other; for example, wetland preservation reduces sea level rise, storms, etc.

Structural or Infrastructural

Adaptation may be institutional, administrative, organizational, regulatory, educational, or financial (incentive and/or subsidies on one hand and taxes on the other hand). Further, adaptation measures may be technological, economical, and legal.

Spatial Scale

The adaptation may be localized or widespread.

Adaptation Function

The adaptation may function differently to adjust climate change, that is, retreat, accommodate, and protect.

5.2.3 Fundamentals of Adaptation

There are four fundamentals of adaptation:

1. Adaptation to what

Climate stimuli for adaptation are expressed as

- weather conditions (average, daily, and hourly);
- climate variability (interseasonal or intraseasonal, long-term, decadal, or catastrophic extreme events);
- climate change (isolated or recurring and gradual or sudden).

In addition to climate, other intervening conditions (economic and institutional arrangements) and type of effect also influence the sensitivity of the system. The types of effect are

- adverse (vulnerability);
- risks and perceptions of risk;
- ecological (drought-magnitude or frequency);
- crop failure or income loss;
- past, actual, or anticipatory changes;
- sensitivity and adaptability of the system (high or low and fast or slow).

Recent emphasis is given on system-relevant climate-related stimuli by examining sensitivity of the system rather than by considering only the limited array of variables provided in the climate change scenario.

2. Who or what adapts

The adaptation may be made by crop species, crop, and cropping and ecological system for

- vulnerability (degree to which a system is susceptible to injury, damage, or harm);
- variability (variation in climate parameter in a given period having the same mean value);
- sensitivity (degree to which a system is affected by or responsive to climate stimuli);
- susceptibility (similar to sensitivity, with some connotations toward damage);
- resilience (degree to which a system rebounds, recoups, or recovers from stimuli);
- flexibility (degree to which a system is flexible or accommodating).

3. How does adaptation occur?

It means which process will be modified in the system to alter climate impact or resulting outcome or condition, for example, altering the phenological development of crops by changing the planting date.

4. How good is the adaptation?

Goodness of adaptation is evaluated based on costs, benefits, efficiency, and implementability from cost-benefit ratio and adaptation efficiency. Cost of adaptation should be less than the additional benefit, depending upon the nature of climate change and level of the impact. In addition to cost, it is also important to include social welfare and equity.

5.2.4 Areas of Adaptation

There are four areas of adaptation to minimize climate change impacts (Matthews, Rivington, Muhammed, Newton, & Hallett, 2013):

1. Introduction of new crops

To introduce new crops in the changed climate, previous crops that are not sustaining yield due to more erratic weather, faster crop maturation, less water availability, or the invasion of new pests and diseases are to be phased out. For that, one has to

- identify "analog climates," which would give indications as to which crops might grow where or other possible alternatives in the future for a particular location;
- assess temperature- and P_{cp}-driven agrometeorologic metrics (AMM) like the start of growing season, plant heat-stress days, growing degree days, heat wave days, cold and dry spells, dry and wet days, soil moisture, and water deficit days for the changed climate;
- identify movement of pests and diseases with shifts in crop distribution;
- promote urban agriculture and identify the best management practices and trade-offs compared with rural agriculture,
- explore the possibility to bring current marginal land into production.

Although phasing out of previous crops or introducing new crops with climate change is the best option, some constraints such as geographic barriers imposed by terrain, other land uses like forests and wetlands, trade-offs between crop production, and other ecosystem services can limit this area of adaptation.

2. Development of new varieties of the existing crops

The increasing temperature in the future demands the development of heat tolerance in plants, which can be created by (i) selecting and breeding

heat tolerance in crops, for example, in wheat by crossbreeding wheat (*Triticum aestivum*) with its key ancestors, *Aegilops tauschii* and *T. durum*; (ii) introducing antioxidant defense system; (iii) increasing heat shocking proteins; and (iv) developing stay-green genotypes (which maintain leaf chlorophyll and photosynthetic capacity at high temperature). In this area, genomics research plays an important role in modifying plant functions like
- transferring C4 pathways into C3 species to increase photosynthesis rate;
- nitrogen (N) fixation into cereals;
- resistance to pests, diseases, drought, heat, and salinity through hybrid vigor.

To bring out the resistances, genetic modification techniques of traditional breeding and advanced biotechnology such as genetically modified crops with pest resistance (Bt) and herbicide tolerance (Roundup Ready) and marker-assisted selection are to be adopted. The emerging field of proteomics is of help in this direction, which relates information at the genotype level to that at the phenotype. Crop modeling can also support to understand genotype-environment interaction for selecting varieties for different environments. With many of these modifications that help in mitigation and adaptation to climate change, there are some trade-offs to be made. For instance,
- increasing photosynthetic rates may lower the protein contents;
- N fixation in cereals and resistance to pests and diseases may affect the plant's energy budget;
- breeding for drought resistance by selecting deeper rooting characteristics may divert biomass away from the shoots and yield component, resulting in lesser yield in good years;
- breeding crops with improved N uptake ability will be limited by the amount of N available in the soil;
- enhancing N use efficiency affects nutritive value adversely by lowering protein content.

Often, adaptations at the plant level do not go through to farm level due to constraints like resource availability; soil constraints; menace of pests, diseases, and weeds; and socioeconomic factors. For example, deep rooting trait developed for drought resistance may not flourish under high soil strength and P_{cp} instability due to restricted root proliferation.

3. Evolution of crop management practices

Under increased temperatures, shortening of crop duration and faster depletion of soil water by accelerating evapotranspiration rates reduce crop yield. Crops may also face nutrient shortages and experience different and more intense pest and disease pressures under the changed climate. Under such conditions, interactive effects between CO_2, temperature, and P_{cp} on crop production are likely to be important. Therefore,

- introduce diversification; if one crop does not do well or fails, the other can benefit and compensate;
- grow more crops instead of one under favorable longer-growing season;
- introduce heterogeneity by intercropping and agroforestry to build resilience to environmental change into cropping systems;
- adjust planting date so that crop growth period synchronizes the favorable weather conditions;
- improve irrigation efficiency by replacing flood irrigation with micro-irrigation system such as drip in areas of adequate access to irrigation water, well-timed "deficit irrigation" in areas with limited access to irrigation, and water conservation and water harvesting techniques in nonirrigated areas;
- improve fertilizer efficiency by adopting technology of need-based fertilization like leaf color chart;
- minimize relying on artificial fertilizers and agrochemicals to have benefit of reduced GHG emissions; use biological pest control;
- adopt organic farming and conservation agriculture;
- use integrated approaches of biological processes by encouraging beneficial soil microbial populations for more efficient nutrient cycling;
- adopt proper crop density (number of plants m^{-2}) in the field;
- adopt soil-specific adaptation, particularly for water and fertilizer management.

Though there are a number of adaptation options, but the selected adaptation must fulfill the criterion of economic, infrastructure, institutional, and cultural aspects and variability in local conditions. Moreover, the adaptation options are very complex, and crop models have been used extensively to clarify this complexity, that is, planting dates, varieties, irrigation, and fertilizer. For example, CropSyst model has made it possible to find out the best management practices for rice-wheat system in Indian Punjab, that is, irrigation scheduling based on pan evaporation; planting dates of rice and wheat as 171 and 308 Julian day, respectively, and $120\,kg\,ha^{-1}$ fertilizer for each crop at present time slice (Jalota et al., 2011); and shifting planting dates to 185 for rice and 322 Julian day for wheat as an adaptation measure (Jalota et al., 2012; Jalota, Kaur, Kaur, & Vashisht, 2013; Jalota, Vashisht, Kaur, Kaur, & Kaur, 2014) for time segments of 2021–50 and 2071–98, respectively. Another study by Jalota and Vashisht (2016) portrays that in the future years (2020, 2050, and 2080), shifting transplanting of maize by 7 days (from 189 to 196 Julian day) in subhumid zone, rice by 14 days (from 171 to 185 Julian day) in semiarid zone, and cotton by 21 days (from 121 to 142 Julian day) in arid zone emerges as doable adaptation measure to minimize yield reduction of crops (Table 5.1).

Similarly, Vashisht et al. (2013) using CERES-Wheat model concluded that planting of rain-fed wheat up to 329 Julian day in Northern India could combat climate change impact till 2030, but not beyond that. Logically, by shifting planting date, plant growth period gets synchronized with favorable weather conditions and improves crop yields. However, the benefit of shifting for improving yield is more in the years of adverse weather and less in the years of favorable weather (Jalota et al., 2013). For example, under adverse years when yields were low, shifting transplanting date of rice from Julian day 171 (normal transplanting date) to 178, 185, and 192 Julian day improved the yield by 3%, 11%, and 12% in the midcentury (MC, 2021–50) and 3%, 14%, and 18% in the end century (EC, 2071–98), respectively (Table 5.2).

Table 5.1 Projected Percent Reduction in Yield of the Three Cropping Sequences as Influenced by Shifting Transplanting Dates in the Future

	2020	2050	2080	2020	2050	2080
	Maize			**Wheat**		
NPD	−1.2	3.7	15.0	1.5	7.4	21.4
NPD + 7 days	2.8	6.5	16.4	27.9	11.8	17.4
NPD + 14 days	7.6	12.6	20.7	26.4	32.6	4.7
NPD + 21 days	13.2	18.6	26.2	15.2	25.5	32.2
	Rice			**Wheat**		
NPD	2.5	5.3	12.1	5.0	0.6	−0.1
NPD + 7 days	0.4	2.4	9.2	0.9	21.6	11.9
NPD + 14 days	1.2	2.4	8.8	8.2	8.2	10.4
NPD + 21 days	2.9	4.3	10.1	4.1	1.4	1.9
	Cotton			**Wheat**		
NPD	28.0	48.7	66.4	−0.9	2.7	9.5
NPD + 7 days	20.2	41.8	61.9	15.9	10.7	13.6
NPD + 14 days	5.8	27.3	51.5	−9.9	26.3	2.7
NPD + 21 days	−4.8	17.2	42.6	6.1	12.9	19.6

Values with negative sign indicate increase in yield. NPD represents normal planting date, which is 189, 171, and 121 Julian day for maize, rice, and cotton crops, respectively.
Data from Jalota, S.K., & Vashisht, B.B. (2016). Adapting cropping systems to future climate change scenario in three agro-climatic zones of Punjab, India. *Journal of Agrometeorology, 18*, 48–56.

Table 5.2 Improvement in Crop Yields by Shifting Transplanting Date of Rice and Wheat in Different Time Slices of the 21st Century

Yield (kg ha^{-1})	In High-Yield Years		In Low-Yield Years	
	Midcentury	End Century	Midcentury	End Century
Rice				
NPD	6856	6624	6109	4769	3855
NPD + 7 days	7068 (+3%)	6627 (+0%)	6240 (−2%)	4893 (+3%)	3974 (+3%)
NPD + 14 days	7295 (+6%)	6517 (−2%)	5959 (−3%)	5314 (+11%)	4384 (+14%)
NPD + 21 days	–	6449 (−3%)	5863 (−4%)	5364 (+12%)	4565 (+18%)
Wheat				
NPD	5790	7207	7931	2184	2026
NPD + 7 days	5940 (+3%)	7723 (+7%)	7756 (−2%)	3937 (+80%)	2979 (+47%)
NPD + 14 days	5730 (−1%)	7404 (+3%)	7588 (−4%)	4944 (+126%)	4044 (+100%)
NPD + 21 days	5721 (−1%)	7417 (+3%)	7778 (−2%)	5245 (+140%)	4602 (+127%)

NPD is normal transplanting date, which is 121 Julian day for rice and 319 for wheat. Values in the bracket are percent change. Midcentury and end century represent years 2021–50 and 2071–98, respectively.
Data from Jalota, S.K., Kaur, H., Kaur, S., & Vashisht, B.B. (2013). Impact of climate change scenario on yield, water and nitrogen—balance and—use efficiency of rice-wheat cropping system. *Agricultural Water Management, 116,* 29–38.

The enhancement in wheat yield was 80%, 126%, and 140% in MC and 47%, 100%, and 127% in EC by changing the date of wheat planting from Julian day 309 (normal transplanting date) to 316, 323, and 330, respectively. Another adaptation option for ameliorating the adverse effect of increased temperature is keeping the soil moist with frequent irrigations or covering the soil surface with vegetative mulches. Frequent irrigations or straw mulching increases the proportion of water in the soil-water mixture and reduces temperature by increasing the specific heat of the soil-water mixture (Jalota & Prihar, 1998). The effect of temperature reduction with mulching is large, rapid, and short-lived in the surface layer and small, slow, and long-lived in the deeper layers (Singh & Sandhu, 1978). Mulching with crop residues usually reduces the maximum soil temperature because of greater reflection and less absorption of solar radiation and lower thermal conductivity of the mulch than that of the soil. Lower temperature underneath the mulch lowers the vapor pressure and, thus, the vapor pressure gradient between the soil surface and the atmosphere. It also decreases wind turbulence and increases resistance to vapor flow and conserves soil moisture. Straw mulching also combats the impact of increased temperature by enhancing availability of N through mineralization and reducing transpiration from weeds suppressed through the smothering effect. The beneficial effect of straw mulching under summer crops (high temperature) in terms of increasing yield, saving of water, and N inputs in different crops has been documented. Jalota, Khera, Arora, and Beri (2007) worked out the economic feasibility of this practice in North India considering benefit-cost ratio and found that the practice has the economic viability for mentha, rain-fed maize and sugarcane, irrigated maize in loamy sand soil, sunflower, winter maize, potato, and soybean, when straw mulching @ $3\,t\,ha^{-1}$ cost (C1) is $58.8\,ha^{-1}$ (Table 5.3). However, mulching @ $6\,t\,ha^{-1}$ costing $103\,ha^{-1}$ (C2) is economical for mentha and maize crops only.

Table 5.3 Economic Return of Straw Mulching in Different Crops

Crop	Yield Increase $(t\,ha^{-1})$	Saving		Gross Return $(\$\,ha^{-1})$				Benefit-Cost Ratio	
		Irrigation Water (cm)	N (kg)	Yield	N fertilize	Water	Total	B/C1	B/C2
Maize fodder	7.5	15	50	44.1	7.4	7.4	58.8	1.00	0.57
Soybean	0.4	–	–	88.5			88.5	1.51	0.86
Sorghum fodder	7.2	23	50	31.8	7.4	11.0	50.1	0.85	0.49
Mentha	0.7	32	25	216.2	3.7	18.4	238.2	4.05	2.31
Sugarcane	4.3	40	–	56.9		22.1	79.0	1.34	0.77
Potato	3.9	12	–	86.0		7.4	93.4	1.59	0.91
Moong	0.1	7	–	20.7		3.7	24.4	0.42	0.24
Winter maize	1.0	23	–	80.9		11.0	91.9	1.56	0.89
Maize rain-fed	1.1	–	–	97.6			97.6	2.65	1.66
Maize (irrigated, loamy sand)	1.9	–	–	173.2			173.2	2.95	1.68
Maize (irrigated, sandy loam)	0.4	–	–	39.2			39.2	0.67	0.38
Sunflower (loamy sand)	0.4	–	–	86.0			86.0	1.46	0.84
Sunflower (sandy loam)	0.2	–	–	52.9			52.9	0.90	0.51

B/C is benefit-cost ratio. C1 and C2 are the costs of mulching at 3 and $6\,t\,ha^{-1}$, respectively.
Modified from Jalota, S.K., Khera, R., Arora, V.K., & Beri V. (2007). Benefits of straw mulching in crop production. *Journal of Research (PAU), 44,* 104–107.

4. Risk assessment of annual weather variability and extreme events
Although in the climate change projections there are a number of uncertainties yet advances in climate science, meteorology, and global data observation and monitoring networks in the recent years have enabled the reasonably accurate prediction or real time forecast of seasonal climate up to a 3-month lead time.

The knowledge of reliable forecasts, environmental changes, and their spatial and temporal magnitudes serves important role in reducing risks of extreme weather and climate variability during the cropping period. For assessing the crop response to risks and defining optimal agronomic practices under such conditions, a combination approach of field work and simulation modeling could be applied. The combination approach (Jalota et al., 2011, 2012) and simulation using CropSyst model (Jalota et al., 2013, 2014; Jalota and Vashisht, 2016) were able to furnish suitable dates of transplanting (where risk of crop yield loss is less) rice–wheat, maize–wheat, and cotton–wheat cropping systems in the three agroclimatic zones of Punjab in different time slices of the 21st century using projected climate data on CO_2, temperature, and P_{cp} from GCM and RCM models. Similarly, Nandel, Kerusebaum, Mirschel, and Wenkel (2012) using MONICA model and Vashisht et al. (2015) using DSSAT recommended adaptation options for sugar beet in Germany and for potato in Minnesota state of United States, respectively. Another adaptation for risk management is crop insurance.

5.2.5 Potential Adaptations

There are a number of potential adaptations that if adopted singly or in combination not only will offset the negative climate change impacts but also will take advantage of the positive ones. The potential adaptations are
- altering/changing the timing or location of cropping activities;
- identifying varieties/species with more appropriate thermal time and vernalization requirements and/or with increased resistance to heat shock and drought;
- developing and using crop varieties and species resistant to pests and diseases;
- altering fertilizer rates to maintain the grain or fruit quality consistent with the prevailing climate;
- altering amounts and timing of irrigation and other water managements;
- wider use of technologies to harvest water and conserve soil moisture where P_{cp} decreases;

- managing water to prevent water logging, erosion, and nutrient leaching where P_{cp} increases;
- diversifying farming activities to enhance income through integration with others such as livestock rearing;
- improving the effectiveness of pest, disease, and weed management practices through wider use of integrated pest and pathogen management;
- using climate forecasting to reduce production risk.

The benefits of adaptation vary with crop, temperature, P_{cp}, and landscape changes (Easterling et al., 2007; Meyer & Rannow, 2013). The benefits are more for wheat than rice and maize, and the benefit for wheat crop is almost the same in temperate and tropical system. As climate changes, its effects are limited not only to direct changes of climate conditions and crop productivity but also on landscapes and their services provided for society. Therefore, for the development of effective and efficient adaptation strategies of specified landscape ecology, it is essential to realize the complex interaction of abiotic, biotic, and socioeconomic systems and its effects on land use or even ecosystem services. Further, the focus of adaptation should be on strengthening the main issues like spatial planning, GHG mitigation, food security, water management, hazard prevention, or conservation of biodiversity. Thus, successful adaptation and management need to be geared toward the maintenance of environmental integrity, the protection against unwanted or even hazardous processes, and the exploitation of beneficial opportunities in response to actual or expected change. In addition, it is essential to measure the success of adaptation activities and the trade-offs between different adaptation options (e.g., hard and soft flood protection).

5.2.6 Agrometeorological Metrics for Adaptation

Potential climate change values in terms of summaries (i.e., annual or mean monthly values) are of less importance to infer adaptations. Therefore, methods are to be developed for communicating climate change; it is likely impacts and decisions to control. One method of communication is agrometeorologic metrics (AMM). AMM are defined as values that describe a property, either of the climate itself or an entity or process that is affected by it. AMM indications are very important for determining how changes in the biophysical environment can lead to land management and policy adaptations to achieve the objectives. AMM, when derived from future climate projection data from climate models, provide important indications of future bioclimatic conditions within which agriculture, forestry, conservation, and many other land uses will have to operate. The AMM can be either quantitative (i.e., trends) or qualified (i.e., good, bad, and neutral) (Rivington et al., 2013).

Based on the source by which AMM is influenced, it can be classified as (i) rain-driven (soil moisture, days and period of soil moisture deficit, dry-soil days, the date for the end of field capacity, reaching the maximum soil moisture deficit, and return to field capacity) and (ii) temperature-driven (start of growing season, plant heat-stress days, and growing degree days) (Table 5.4). The indicators used for AMM are date, count, and threshold values and indices.

In summary, AMM provide the detailed information about the changes in climatic and climate-soil-related constraints at particular locations needed to make specific choices (strategic or possible tactical interventions).

5.2.7 Barriers in Adaptation

The barriers in adaptation are
- not seen as a big problem yet, so the temptation is to wait for the impact and then react;
- uncertainty of climate information, belief that the uncertainty is too great to warrant taking adaptation action now;
- the lack of useful precedents or evidence of adaptation actions, what are others doing;
- the lack of availability or restricted access to appropriate technologies;
- the lack of availability of resources such as in-house expertise;
- the lack of supportive policies, standards, regulations, and design guidance, encouraging status quo and/or presenting impediments to progress;
- limited budgets to meet the costs of identified adaptation options;
- social/cultural/financial rigidity and conflicts;
- limited understanding of climate risks and vulnerabilities at the present and in the future;
- existing legal or regulatory restrictions;
- the lack of acceptance/understanding of risks associated with implementation due to the lack of decision or ineffective leadership;
- relevance of climate information for development-related decisions;
- trade-offs between climate and development;
- inadequate resources; resources prove to be important in almost every stage, but certainly in implementation and monitoring phases of adaptation;

Table 5.4 Some Important Agrometeorologic Metrics, Their Units, and Definitions

Type	Metric	Unit	Definition
Date	Start of growing season	Julian	Day when five consecutive days $T_{ave} > 5.6°C$ (from January 1)[a]
	Date of start of max soil moisture deficit	Days	Day when SMD is at maximum[b]
	Wettest week		Midweek date, when max 7-day value of P_{cp} occurs[a]
	Return to field capacity		Days when soil moisture deficit <5 mm (after date of max SMD[b])
Count	Growing season length	Days	Days when $T_{ave} > 5.6°C$ between start and end of growing season[b]
	Dry		Days when $P_{cp} < 0.2$ mm[a]
	Wet		Days when $P_{cp} > 0.2$ mm[a]
	Plant heat stress		Days when $T_{max} > 25.0°C$[a]
	Dry-soil days		When soil moisture is less than permanent wilting point[b]
	Very dry-soil days		When soil moisture is less than air-dry soil[b]
Degree days	Growing degree days	Day	$\sum T_{ave} > 5.6°C$[a]
	Heating degree days	deg	$\sum 15.5°C - T_{ave}$ where $T_{ave} > 15.5°C$[a]
Water	Excess winter rainfall	mm	$\sum P_{cp} >$ saturation capacity[a]
	Wettest week amount		Maximum amount of P_{cp} (seven consecutive days)[a]
	Minimum soil water		Max soil moisture deficit[b]
Wave	Heat wave	Days	Max count of consecutive days when $T_{max} >$ average T_{max} (baseline year) $+3.0°C$ (minimum of 6 days)[a]
	Cold spell		Max count of consecutive days when $T_{min} <$ average T_{min} (baseline year) $-3.0°C$ (minimum of 6 days)[a]
	Dry spell		Maximum consecutive count, $P_{cp} < 0.2$ mm[a]
	Wet spell		Maximum consecutive count, $P_{cp} > 0.2$ mm[a]
Indices	P intensity	Index	$\sum P_{cp} > 0.2$ mm/count day $P_{cp} > 0.2$ mm[a]
	P seasonality		$S =$ winter, $P =$ summer, P_{cp}/total P_{cp}[a]
	P heterogeneity		Multiple Fournier index

[a]Single-variable-based.
[b]Multiple-variable-based; SMD is soil moisture deficit; P_{cp} is rainfall; T_{max}, T_{min}, and T_{ave} are maximum, minimum, and average temperature, respectively.
Modified from Rivington, M., Matthews, K.B., Buchen, K., Miller, D.G., Bellocchi, G., & Russell, G. (2013). Climate change impacts and adaptation scope for agriculture indicated by agro-meteorological metrics. *Agricultural Systems, 114*, 15–31.

— information and communication about the problem;
— values and beliefs that influence how people perceive, interpret, and think about the risks and their management.

Adaptive capacity is enhanced if these barriers are removed by improving the knowledge of climate change and associated risks and vulnerabilities along with knowledge and updating of institutional and legal frameworks.

5.3 MITIGATION OF GREENHOUSE GASES EMISSION

The emissions of N_2O, CO_2, and CH_4 gases from cropped soils not only are the loss of the nutrients needed for plant growth but also cause global warming as discussed in Chapter 1. The opportunities for mitigating GHGs in agriculture center around three basic principles, that is, (i) reducing emissions; (ii) enhancing sink or removals; and (iii) avoiding or displacing emissions. The following are a number of strategies to mitigate emission of GHGs.

5.3.1 Strategies to Mitigate Nitrous Oxide Emission

Soil surface N fluxes are influenced by the form and quantity of added N. On average, 0.2%–1.5% of applied N to agricultural soils is emitted as N_2O (Groot Jeroen, Walter, Egbert, & Lantinga, 2006) and the remaining losses as nitrate (NO_3^-) leaching, erosion, and denitrification. Since a substantial proportion of the GHGs produced by agriculture is attributable to the production and application of N fertilizer alone (Stern, 2006), breakthroughs in N use efficiency could substantially mitigate emissions in agriculture. The emission of N_2O is regulated by climatic (CO_2, temperature, and P_{cp}) and soil (water content, water table depth, OC, and pH) factors and management interventions. These factors control both nitrification and denitrification rates and the ratio of $N_2O:N_2$ from denitrification and ultimately N_2O emission. These factors and processes can be regulated by the following managements.

Improving Water Management

It is well known that maximum denitrification rates (chemical reactions that produce N_2O) occur when soil-water-filled pores are $\geq 70\%$ (low–oxygen conditions), but at this water content, chances of N leaching losses are more. So any water management practice that optimizes water use by avoiding excess irrigation and maintains a balance between limiting denitrification and NO_3^- leaching is desirable. Drip-irrigated systems are an example of such water management practice (Kallenbach, Rolston, & Horwath, 2010) that distributes applied water and N uniformly to the area of maximum crop uptake and minimizes leaching and denitrification. In sprinkler, furrow, or flood irrigation systems that reduce N_2O emission along with irrigation water, N application rates are matched as per plant requirement. Application of fertilizer immediately after P_{cp} also increases N use efficiency of plants and mitigates N_2O emissions. Intermittent irrigation in rice though reduces CH_4 substantially increases the emission of N_2O, an even more potent GHG than CH_4. This complex relationship requires site-specific information, which can be accomplished by field studies for a wider range of locations. However, simulation models may help looking at prior assessments of management impacts on mitigation of GHG emissions.

Minimizing Bare Soil
Keeping bare fallows between cropping seasons is a common practice. During that period, NO_3^- accumulates as a result of mineralization of OM and nitrification of NH_4^+. But these NO_3^- are more susceptible to loss by denitrification without contributing to overall productivity than that which is produced when plants are available. Recent opinions to avoid N_2O emissions in bare soils are system intensification and cultivation of cover crops during fallow. System intensification reduces N_2O emissions by increasing N retention in cropping systems and leaving less for N_2O emission. Cover crops reduce indirect N_2O emissions through (i) immobilizing N; (ii) fixing N biologically; and (iii) protecting soil against erosion. Their (cover crops) effect on direct N_2O emissions is site-specific; for example, N_2O emission rates are identical in legume cropping and bare soils in the Mediterranean semiarid environment of Western Australia (Barton, Butterbach-Bahl, Kiese, & Murphy, 2011) and higher in California due to lower C:N ratio and shallower roots, which could not extract soil N more efficiently (Kallenbach et al., 2010). This calls for the understanding of the effect of mixture of legumes and nonlegumes to get more benefits of biological N fixation.

Tightening the N Cycle Through N Immobilization
Incorporation of pruned residues of woody crops (lignin- and polyphenol-rich materials) into the soil reduces N_2O emission due to their strong N immobilization effect by enhancing C sequestration, creating biodiversity (Holtz & Caesar-Ton That, 2004), and reducing soil erosion (Rodriguez-Lizana, Espejo-Perez, Gonzalez-Fernandez, & Ordonez-Fernandez, 2008).

Use of Nitrification Inhibitors

Nitrification inhibitors reduce N_2O emission directly by reducing nitrification and indirectly by reducing the availability of NO_3^- for denitrification. Nitrification and urease inhibitors are generally used to reduce the N_2O emission. Nitrification inhibitors are used to inhibit nitrification of NH_4^+ by soil bacteria, which slow the microbial processes that lead to N_2O formation. Urease inhibitors are used to reduce the hydrolysis of urea to NH_4^+. For inhibitors to be successful, the criteria to be fulfilled are (i) the activity of nitrification must be inhibited until fertilizer penetrates the soil surface or mineral N derived from it is assimilated by plants; (ii) gaseous emission in the field must be reduced; (iii) crop or forage yield must be increased; (iv) the inhibitor must cost less than the NH_3 volatilization or denitrification; and (v) the inhibitor must be capable of being incorporated with the fertilizer N, be stable during manufacture and storage, and be environmentally safe. The most extensively tested inhibitor is nitrapyrin, which inhibits the first step in nitrification, that is, the oxidation of NH_4^+ to NO_2 by inhibiting the cytochrome oxidation by nitrosomonas. The addition of nitrapyrin results in a decrease and delay of NH_4^+—N disappearance, accumulation of much lower soil NO_3^-—N contents, and a substantial reduction in N_2O emissions (Abbasi et al., 2003). The application of nitrapyrin and dicyandiamide (DCD) to grassland reduced the emission of N_2O from NH_4^+-based fertilizers by 64% and 52%, respectively (McTaggart, Clayton, & Smith, 1994). The use of nitrification inhibitors such as S. benzylisothiouronium butanoate (SBT butanoate) reduced emissions of N_2O by 4%–5% compared with urea treatment alone (Bhatia, Pathak, Aggarwal, & Jain, 2010).

Use of Slow Release/Super Granule/Fertilizers

The recovery efficiency of applied N is about 50%; the rest goes as leaching, runoff, erosion, and gaseous emissions (Ladha, Pathak, Krupnik, Six, & Van Kessel, 2005). Some forms of N fertilizers emit more GHGs than others due to their inefficient use by the plants. For example, anhydrous NH_3 produces the highest N_2O emissions, whereas forms containing NO_3^- produce the lowest. Based on the concept of limiting substrate, slow-release, controlled-release, and stabilized fertilizers reduce losses of N via drainage or atmospheric emission by enhancing crop recovery. Though in these fertilizers sufficient N to satisfy plant requirement is applied in a single application, very low concentration of mineral N remains in the soil throughout the growing season. In urea- and NH_4^+-based fertilizers, release of N can be made slow by a number of ways such as changing the size of fertilizer granules, chemical modifications, and coatings. For example, increasing the size of urea granules from the conventional $0.01-1$ g decreased the nitrification rates (Skiba, Fowler, & Smith, 1997). Super granules decrease NH_3 volatilization even in highly permeable soils. A combination of increasing the size of pellet to 1 g and adding nitrification inhibitor DCD slowed the nitrification rates to the extent that 30% of the original N application persisted up to 2 months after fertilizer application (Goose & Johnson, 1993). SBT fluorate with urea slowed down the release of N and reduced global warming potential by 13.5% and 19.5% compared with urea alone under conventional and zero tillage, respectively (Bhatia, Sasmal, Jain, Kumar, & Singh, 2010). Cost-effective indigenous materials such as neem (*Azadirachta indica*)-coated urea are being recommended to suppress nitrification in semiarid environments (Thind et al., 2010).

Site Specific Nutrient Management

Fertilizers are direct and indirect sources of GHGs, so any increase in fertilizer use efficiency would translate into reduced emissions of both CO_2 (fossil fuel embedded in the fertilizer) and N_2O. An advanced approach to enhance fertilizer use efficiency, that is, optimizing fertilizer application for high yield, is termed site-specific nutrient management (SSNM). SSNM is a technology that can be tailored to local needs both in space and time, ensures a rapid take-up of N by the plants and high yields, reduces residence time of N in the soil, and reduces release of N_2O or other reactive N compounds to the atmosphere or draining water. A research conducted by the International Rice Research Institute (IRRI) and its partners in various countries of Asia depicts that the benefits of SSNM in rice-based cropping systems include increase in rice yields and N use efficiency by 30%–40%. This offers significant potential for decreasing N fertilizer-induced GHG emissions associated in rice-based system. This dual benefit of increased yield and reduced GHG emissions is more with balanced fertilizer application in the SSNM approach. If at any location there is imbalance of fertilizers, that is, phosphorus is deficient, it will not only cause decrease in yield but also stimulate high root exudation and increase CH_4 emissions. In the future under elevated CO_2 concentration, SSNM approach would be of greater use to tap the so-called "CO_2 fertilization" effect under the environment where temperature is not reaching the critical level.

Use of Leaf Color Chart

Appropriate dose of fertilizer and its application coinciding the period of rapid plant uptake are important mitigation strategies to reduce N_2O emission. In that context, practices like leaf color chart (LCC), chlorophyll meter (SPAD), and green-seeker are used, which reduce N_2O emission due to the reason that need-based applied N is taken up by the plants promptly, leaving little in soil for transformations and emission (Bijay-Singh et al., 2012). The LCC is used for real-time N management and synchronizing N application with crop demand. Bhatia, Pathak, Jain, Singh, and Tomer (2012) recorded reduction of N_2O by 16% in rice and 18% in wheat with application of $120 \, kg \, N \, ha^{-1}$ at LCC ≤ 4 as compared with conventional method. They believed that LCC-based urea application can reduce global warming potential of a rice-wheat system by 10.5% in Indo-Gangetic Plains of India. Unfortunately, multiple applications of fertilizer are not practical in many agricultural situations due to P_{cp} patterns and difficulty in applying fertilizer within a maturing crop canopy.

Integrated Nutrient Management

In modern agriculture, high-yielding varieties need more macro- and micronutrients, which are supplied through the combination of inorganic and organic fertilizers. Generally, more than 50% of applied N is not assimilated by plants (Ladha et al., 2005). The increasing dose of N through inorganic fertilizer will cause emission of N_2O with CO_2 emitted during production process. Heavily manured lands also emit a lot of N_2O; therefore, it should be applied only as per need. Hence, it is important to practice integrated use of nutrients from inorganic and organic fertilizers to minimize emissions of N_2O and CO_2. For example, applying urea fertilizer with C source (wheat straw and green manure) substantially reduces N_2O emission compared with urea alone (Aulakh, Khera, Doran, & Bronson, 2001), possibly due to microbial immobilization. Integrated use of nutrients and using crop mixtures also encourage diverse and beneficial soil microbial populations for more efficient nutrient cycling.

Organic Amendments

Organic farming reduces N_2O emission as mineral N is not used as fertilizer, and the soils in organic farming are more aerated and have lower concentrations of mobile N (NO_3^-). Since organic crop systems are limited by the availability of N, their emissions are lower than those of conventional farming systems per unit of land area. However, with lower yield from organic farming, the emissions per unit of produce could be the same or higher (Petersen et al., 2006). A diversified crop rotation with green manure in organic farming improves soil structure and diminishes emission of N_2O, even though the N provided by the green manure does contribute to N_2O emissions. Thus, the substitution of urea by organic sources can be a good management strategy to reduce N_2O emission from soils.

Rate, Source, and Timing of N Fertilizer Application

N_2O emissions can be reduced by using the appropriate rate, source, and timing of fertilizer N. The risk of N_2O emission increases when N is applied above the economic optimum N rate (Pathak, 2010) or when available soil N (especially in NO_3^- form) exceeds crop uptake. Anhydrous ammonia emits N_2O at rates 2–4 times greater than from those amended with urea and ammonium nitrate (Venterea, Burger, & Spokas, 2005). Application of N fertilizer at vegetative phase of crops or placement of fertilizer N into the soil near the zone of active root uptake reduces N_2O losses (CAST, 2004).

5.3.2 Strategies to Mitigate Carbon Dioxide Emission

Reduce the CO_2 Release From Energy Use

Emissions of CO_2 from fossil fuels can be brought under control by using non-CO_2-emitting energy sources such as wind, solar, or nuclear energy or storing the CO_2 generated from fossil fuel, which prevents its release to the atmosphere.

Replace Fossil Fuels With Biofuels

Displacing the use of fossil fuels or decarbonization of fuel use in agriculture sector is one of the strategies to mitigate CO_2 emission. Burning of wood, ethanol, and other fuels from vegetation generates CO_2. But that C is from recently photosynthesized atmospheric CO_2 rather than from fossil C. In effect, it recycles CO_2 rather than introducing new or previously dormant C into active cycling. In case excessive fossil C is used to produce the biofuel, the benefits of biofuels are diminished. There are some crops like sugarcane and maize that can be used for biofuel (ethanol) production to minimize the CO_2 emission from fossil fuel sources. But it is not clear whether these crops can compete successfully with farm, forest, and urban waste products as bioenergy feedstock for maximum ethanol production. Moreover, these require complementary innovations, for example, in cellulosic ethanol production.

Increase the Amount of C Stored in Vegetation and Soil

In soil, input of OM is increased by (i) increasing the primary production in the agroecosystem; (ii) leaving a major proportion of the produced biomass; (iii) importing OM from other ecosystems; and (iv) adding organic material such as crop residues, animal manure, and compost. Such inputs of organic sources improve microaggregation, in which C is protected from decomposition and gets stored (Benbi & Senapati, 2010; Sodhi, Beri, & Benbi, 2009). The increase in storage of C as SOM is called C sequestration (Lal, 2004). The relationship between the application of OM to the soil and the gains in SOC varies depending upon practices like land treatments of planting trees, intensity of tillage on cropland, and restoring grasslands on degraded lands. Indeed, C storage will be increased by the practices that increase productivity or reduce the rate of respiration. Main potential to mitigate CO_2 emission is C sequestration in soil as SOC pool is twice as big as atmospheric C pool (Smith et al., 2008). The amount of soil C sequestered depends upon the accumulation of underground root mass affected by the type of cropping system and longevity of the practice, crop residue management, integrated nutrient, quality and quantity of residues, type of organic and amendment, soil type, moisture and temperature, and climatic conditions (Benbi & Brar, 2009; Benbi, Toor, & Kumar, 2012; Kukal, Rassol, & Benbi, 2009; Rasmussen & Collins, 1991; Singh, Jalota, & Sidhu, 2005; Singh, Jalota, & Yadvinder-Singh, 2007). Now, the interest has been increased in developing methods to sequester atmospheric C, due to more CO_2. The CO_2 sequestration not only stores the C in soil but also reduces the fertilizer N required for obtaining a targeted yield (Benbi & Chand, 2007). The practices adopted for sequestering C in soil are discussed below.

Conservation Tillage

The practices that limit soil disturbance such as conservation tillage (no, reduced, and minimum tillage) or shallow tillage and consequently soil C losses through reduced microbial decomposition are supposed to increase SOC sequestration in cropland soils. In no-tillage (NT) system, C inputs to the soil are more in herbaceous cropping system because straw is retained instead of being removed or burned as compared with woody crops, where herbicides are used to lower weed biomass compared with conventional tillage. In most of the cases, the reduced-tillage (RT) practices do not ensure a persistent organic C sequestration as tillage is resumed to sustain yield, and the fixed C is rapidly lost again from soil. The other shortcomings that limit the practice of RT for soil C sequestration are (i) reduced crop productivity;

(ii) small, inconsistent, and variable C fixation; (iii) temporary sequestration until traditional tillage practice is resumed; and (iv) high N_2O emission. With practices of NT and RT, C sequestration is more due to lower fuel consumption during farm operations. It is estimated that 30 L less consumption of diesel by NT compared with conventional tillage results in reduction of CO_2 emission by $80 \, kg \, ha^{-1} year^{-1}$ (Aggarwal, 2008). Therefore, there is a need to find better alternatives to current soil management practices for SOC sequestration in agriculture. But it requires a long-term application to produce a significant and steady improvement of OC content in cultivated soils.

Agroforestry

Unchecked destruction of forests has increased CO_2 concentration in the atmosphere resulting in global warming. Among several strategies to mitigate CO_2 emission, one of the possible options is agroforestry. Agroforestry is an integrated land–use system combining trees and shrubs with crops and livestock, which helps sequester SOM. The tree in agroforestry enhances SOM by adding above- and belowground inputs to soil (Lal, 2004). Different fractions of SOC like aggregate-associated, labile, hot water-soluble, and particulate C are also higher in agroforestry and grassland systems as compared with agriculture and eroded land (Saha, Nair, Nair, & Kumar, 2010). Under poplar plantation, efficiency of OC sequestration capacity is higher due to conversion of oxidizable C fractions to more of nonoxidizable fraction over a time period of 100 years (Sierra, Martınez, Verde, Martın, & Macıas, 2013).

Fertilizer Management/Integrated Nutrient Management

Fertilizer use sequesters C by increased C input from stimulated biomass production, but such contribution is minor due to complex interactive effects on C transformations in the soil (Ladha, Reddy, Padre, & Van Kessel, 2011). The change in storage corresponds to increase in belowground plant growth. Fertilizers sequester C only if soil carbon storage exceeds the CO_2 emitted. Under N fertilization, soil organic C levels increase only when crop residues are returned to the soil for a long period. For instance, Shang et al. (2011) noticed $470 \, kg \, ha^{-1} year^{-1}$ more C in soil in long-term-fertilized plots than the control in China. At global level, increase in soil C in agricultural ecosystems is 3.5% by addition of N fertilizers (Lu et al., 2012). Biofertilizers build SOM via restoring the soil's natural nutrient cycle with the help of microorganisms.

Organic Amendments

With organic amendments like compost, manures, agroindustrial wastes, and other sources of OM, sequestration of C is high. Compost sequesters more C than manure because of the presence of the more stabilized forms of C. The liquid organic material slurries having very low and easily decomposable amount of C usually do not encourage SOC accumulation. Such types of organic amendments provide more N to microorganisms, which get their added energy requirements through the oxidation of the native organic C.

Residue Incorporation

The effects of crop residue leftover soil surface/applied as mulch on improving soil physical properties, conserving soil moisture, N fertilizer saving, and controlling weeds are well-recognized (Jalota et al., 2007). These crop residues can be incorporated into the soil, which increase soil C (Singh et al., 2005). On decomposition, crop residues supply plant nutrients to the succeeding crop, increase biomass above and below the ground surface, and reduce GHG emission by sequestering soil C. Returning straw to fields rather than burning avoids CH_4 and N_2O emissions and sequesters C (Lu et al., 2010). In real cropping systems, increasing C input is limited due to some constraints such as availability, economical, and social. Availability of crop residue may be less due to nonavailability at the local scale and competition with other uses like feed, fuel, and wood if source is external. The economic constraint is high cost, and social constraint is unawareness of its benefits such as SOC increase and mitigation of GHGs.

Cover Crops

Cover crops increase SOC through their biomass and fixing N from the atmosphere. Green manure crops, termed as cover crops, enhance soil fertility by improving physical and chemical parameters (i.e., soil aggregate stability and soil macro- and micronutrients). These also help in protecting the soil from surface runoff. All these conditions provided by cover crops are favorable for soil C sequestration and reducing CO_2 emission. The common green manure crops are leguminous crops like cowpea, sunhemp, and groundnut. Though legumes sequester C, but their effect is less than cereals due to more decomposition caused by less concentrations of lignin. Due to this reason, cover crops can be combined with the application of material resistant to decomposition like pruning residues of woody crops and agroindustrial wastes to have high potential for C sequestration.

Crop Rotation

By definition, crop rotation is the planned order of specific crops sown on the same field for a period of 2 or more years. The succeeding crop may be of different species (e.g., grain crops followed by legumes) or variety from the previous crop. Crop rotation is an important practice for C sequestration. In fact, rotating to a different crop improves the physical, chemical, and biological environment of soil and reduces pest and diseases, which produce large amounts of biomass and residue for incorporation in the soil for C sequestration.

Organic Farming Systems

Organic farming systems relying on the use of recycled organic materials (e.g., animal manures and compost) offer many benefits, namely, increased SOC storage (sequestration), reduced GHG emissions, lowered energy consumption, and maintained or increased farm profitability because organic farming systems are not allowed to use chemical fertilizers, which consume a large amount of energy for their production. It has been reported that organic farming on the global scale may result in 20% higher SOC sequestration rate (Azeez, 2009) and 40%–60% lower CO_2 emissions (Sayre, 2003) compared with conventional systems.

Pasture Management

Soils under pasture tend to have a higher SOC than cropped soils because they have (i) higher root-to-shoot ratio; (ii) less disturbance; and (iii) lower rates of SOC decomposition. The use of perennial grasses in marginal pastures has the potential to increase SOC stocks as they have deeper, more extensive root system compared with annual ones. A significant increase in SOC stocks of 0.15–$0.35\,\mathrm{MgC\,ha^{-1}\,year^{-1}}$ at the depth from 0 to 20 cm over 6 years in various perennial pasture establishments has been reported (Young, Wilson, Harden, & Bernadri, 2009)

Use of Bt Crops

Crops, varieties, and traits that are resistant to pests and diseases will reduce C emissions due to less number of pesticide applications. Less use of herbicide in such crops can also reduce emissions by enabling farmers to more readily adopt RT or NT systems, which save fuel by reducing the need to plow, adding C to the soil and thereby sequestering C.

Use of Sustainable Land Management Practices

A sustainable use of soil means its exploitation in a way and at a rate that preserves at the long term its multitude of functions and protects or improves its quality while maintaining its potential to meet the likely needs and aspirations of present and future generations (Van Camp et al., 2004). The sustainability of soil system can be maintained by retaining or quickly replenishing OM to avoid degradation. For that, it is crucial to encourage use of sustainable land management (SLM) such as erosion control, water management, and judicious application of fertilizers, which reduce GHG emissions from agricultural lands directly and increase rates of soil C sequestration alternatives to conventional agriculture. Sustainable land management delivers C benefits in three important ways, namely, (i) C conservation in which the large volumes of C stored in natural forests, grasslands, and wetlands remain stored as C stocks; (ii) C sequestration in which the growth of agricultural and natural biomass actively removes C from the atmosphere and stores it in the soil and biomass; and (iii) reduction in emissions of GHGs that emanate from agricultural production.

Use of Biochar

Biochar is produced by controlled pyrolysis (smoldering) of farm residues and can be applied to soils with long-term benefits on soil fertility. It decomposes more slowly and stabilizes biomass C. As compared with conventional crop management, biochar reduces net GHG emissions because of (i) potential energy generation through the combustion process and (ii) C sequestration in the soil. The biochar also has a great potential for improving soil conditions and plant growth by enhancing water and nutrient retention, uptake of applied fertilizers, microbial biomass and activity, and resilience of the cropping systems toward droughts (Lehmann, Gaunt, & Rondon, 2006; Sohi, Lopez-Capel, Krull, & Bol, 2009).

Irrigation Management

Irrigation results either increase or decrease in SOC through its simultaneous effect of increase of net primary productivity and subsequently OM inputs to the soil and soil respiration. Carbon losses occur more under high soil moisture regime, and the best results of C sequestration are more when residues are applied shortly before the onset of the rainy season.

5.3.3 Strategies to Mitigate Methane Emission

Crop Management

Methane emission can be minimized by managing the crops such as (i) growing suitable upland crops to reduce the period of submergence during an annual cropping cycle and diversifying crops such as rice that is replaced with upland crop like maize with less water requirement; (ii) selecting the rice cultivars that transport maximum portion of their photosynthates to panicle growth and grain development rather than growing varieties that use their photosynthates to the development of their vegetative parts (root, leaf sheath, culm etc.); and (iii) selecting cultivars that have minimum tiller numbers and higher proportion of productive tiller, smaller root system, higher root oxidative activity, and higher harvest index. Rice cultivars with few unproductive tillers and transplanting with relatively aged seedlings also mitigate CH_4 emission from rice fields. The rice cultivars *Gayatri* and *Tulsi* developed in India emit 13% and 22% less CH_4 than IR72 (Adhya et al., 2000).

Proper Use of Chemical Fertilizers

The source, mode, and rate of application of mineral fertilizers influence the CH_4 production and emission from rice fields. In soil, N fertilizers influence the activities of methanotrophic microorganisms and inhibit CH_4 emission. There are three mechanisms by which N fertilizers inhibit CH_4 emission, that is, (i) immediate inhibition of methanotrophic enzyme system; (ii) secondary inhibition through the N_2O production from methanotrophic NH_4 oxidation; and (iii) modifying microbial communities in soil. Surface application of SO_4^{2+}-containing N fertilizers is known to reduce CH_4 emission from rice fields by inhibiting CH_4-oxidizing activity in rhizosphere and favoring the activity of other microbial groups over that of methanogens. The foliar spray of N fertilizers is another potential practice for CH_4 mitigation. Addition of N fertilizers in flooded rice soils suppresses CH_4 omission due to the change in redox potential of soil. A mixture of prilled urea and nimin (nitrification inhibitor) also minimizes CH_4 emission. Application of muriate of potash in flooded rice reduces CH_4 emission as it prevents the drop in redox potential and reduces contents of active reducing substances and Fe^{2+} content in rhizosphere. Addition of phosphorus (P) also reduces CH_4 emission as P-deficient soils trigger high root exudates and emit CH_4 more. Nitrification inhibitors such as acetylene, nitrapyrin, thiourea, sodium thiosulfate, calcium carbide, and dicyandiamide waxcoated calcium carbide can also be used for reducing CH_4 emission.

Type of Organic Matter

In integrated nutrient management, organic amendments constitute an important component, but pose problems of CH_4 emission. Addition of fresh organic sources (leguminous green manure and rice straw) to rice soil increases the availability of methanogenic substrates and thereby enhances CH_4 production and emission. However, compost is the most effective for mitigating CH_4 emissions from rice fields.

Water Management in Rice

Reduction in CH_4 emission from agriculture can, to a large extent, be accomplished by changing a rice production system from anaerobic to aerobic and management interventions (alternate wetting and drying of rice field, planting rice on beds, using surface or subsurface microirrigation practices, etc.). Irrigation water should be applied after the soils have dried and fine cracks started appearing. This not only reduces amounts of water application but also reduces CH_4 emissions. Growing of crops on raised beds has shown that there is definite advantage in irrigation water-saving and in reducing GHG emissions though there is no yield advantage compared with flat beds and other cultivation methods (Saharawat et al., 2012). On contrary, there are some reports that support that higher water levels result in lower soil temperature and subsequently CH_4 emissions because transport and release of CH_4 are slowed down by the submergence of aerial parts of rice plants (Neue, 1993). A key problem in modifications of the water management in rice for reducing CH_4 emission is that it concomitantly enhances the emission of N_2O.

Pesticides

In modern crop production, especially in rice, pesticides are increasing. Even recommended doses of most pesticides may affect inhibition or stimulation of certain microbial transformations in rice fields. The pesticides like bromoxynil, methomyl, and nitrification inhibitor (nitrapyrin) are inhibitory to CH_4 oxidation (Topp, 1993). The degree of inhibition of CH_4 oxidation increased with increasing level of 2,4-D with 100 mg 2,4-D/g soil being completely inhibitory.

Improve Digestibility of Fodder by Ruminant

Methane mitigation strategies in ruminants can be broadly divided into preventive and "end-of-pipe" options. The *preventive measures* reduce C and N inputs into the animal husbandry, generally through dietary manipulation.

Alternatively, *"end-of-pipe"* options reduce or inhibit the production of CH_4 (methanogenesis) within the animal husbandry. The loss of methane from ruminant livestock is a problem not only in the respect of GHG emissions but also to farmers. Food converted into and released as CH_4 is not being converted into meat and/or milk. Some strategies to reduce CH_4 in ruminants are the following:

− Introduce methane inhibitors, both biological and chemical, with the animal feed to kill off or at least reduce the activity of the methanogenic microorganisms in the gut.
− Change dietary practices like the addition of ionophores and fats, the use of high-quality forages, and the increased use of grains, which reduce CH_4 through the manipulation of ruminal fermentation and direct inhibition of the methanogens and protozoa or by a redirection of hydrogen ions away from the methanogens or suppressing acetate production, which results in reducing the amount of hydrogen released. Dietary fats reduce CH_4 through biohydration of unsaturated fatty acids, enhanced propionic acid production, and protozoal inhibition.
− Increase the efficiency of nutrients to produce milk or meat, which results in reduced CH_4 emissions. This can be accomplished by feeding high-quality, highly digestible forages or grains.
− Remove protozoa to reduce CH_4 emissions.
− Encourage growth of acetogenic bacteria for removing hydrogen and converting CO_2 and hydrogen to acetate, which the animal can use as an energy source.
− Develop a vaccine that stimulates antibodies in the animal that are active in the rumen against methanogens.

EXERCISES

1. Why adaptation is becoming popular to combat climate change impact? What are the potential adaptations and barriers in adaptation?
2. Under which climatic condition site-specific nutrient management approach would be of greater use to tap the so-called "CO_2 fertilization" effect?
3. Which are the important areas of adaptation?
4. What are the mitigation strategies to reduce the emission of nitrous oxide, carbon dioxide, and methane gases?
5. Which are the water, fertilizer, and crop management practices that help reduce the N_2O emission in rice?

6. Write the mechanisms by which crops and N fertilizer managements affect mitigation of the CH_4.

7. What are sustainable land management practices and in how many ways does it help mitigate CO_2 effect?

8. Fill in the blanks

  (i) N_2O emissions in organic farming are lower than those of conventional farming systems per unit of _____ and higher per unit of _____.

Answer (*land area* and *produce*)

  (ii) The risk of N_2O emissions increases when N is applied above the _____ or when available soil N (especially in NO_3^- form) exceeds _____.

Answer (*economic optimum N rate* and *crop uptake*)

  (iii) Applying urea fertilizer with C source substantially reduces N_2O emission compared with urea alone possibly due to _____.

Answer (*microbial immobilization*)

  (iv) Though legumes sequester C, their effect is _____ than cereals.

Answer (*less*)

  (v) SOC pool is _____ as big as atmospheric C pool.

Answer (*twice*)

  (vi) No tillage does not sequester C in soil in the _____ run.

Answer (*long*)

  (vii) Carbon sequestration gain is higher with compost than with raw manure due to the _____ stabilized forms of C present in the former.

Answer (*more*)

  (viii) The three constraints limiting the intensification of C input in real cropping systems are _____, _____, and _____.

Answer (*availability, economical, and social*)

  (ix) The best results of C sequestration are obtained when residues are applied _____ the beginning of the rainy season.

Answer (*shortly before*)

  (x) A key problem in modifications of the water management in rice for reducing methane emission is that it concomitantly enhances the emission of _____.

Answer (N_2O)

REFERENCES

Abbasi, I., Branyburg, A., Camposponce, M., Abdel Hafez, S. K., Raoul, F., & Craig, P. S. (2003). Diagnosis of *Echinococcus granulosus* infection in dogs by amplification of a newly identified repeated DNA sequence. *American Journal of Tropical Medicine and Hygiena, 69,* 324–330.

Adhya, T. K., Bharti, K., Mohanty, S. R., Ramakrishnan, B., Rao, V. R., Sethunathan, N., et al. (2000). Methane emission from rice fields at Cuttack, India. *Nutrient Cycling in Agroecosystems, 58,* 95–106.

Aggarwal, P. K. (2008). Global climate change and Indian agriculture: Impacts, adaptation and mitigation. *Indian Journal of Agricultural Sciences, 78,* 911–919.

Aulakh, M. S., Khera, T. S., Doran, J. W., & Bronson, K. F. (2001). Denitrifiction, N_2O and CO_2 fluxes in rice, wheat cropping system as affected by crop residue, fertilizer N and legume green manure. *Biology and Fertility of Soils, 34,* 375–389.

Azeez, G. (2009). *Soil C and organic farming: A review of the evidence on the relationship between agriculture and soil C sequestration, and how organic farming contributes to climate change mitigation and adaptation.* United Kingdom: Soil Association. http://www.soilassociation.org/Whyorganic/climatefriendlyfoodsarming/SoilC/tabid/574/Default.aspx.

Barton, L., Butterbach-Bahl, K., Kiese, R., & Murphy, D. V. (2011). Nitrous oxide fluxes from a grain-legume crop (narrow-leafed *lupin*) grown in a semiarid climate. *Global Change Biology, 17,* 1153–1166.

Benbi, D. K., & Brar, J. S. (2009). A 25 year record of carbon sequestration and soil properties in intensive agriculture. *Agronomy for Sustainable Development, 29,* 257–265.

Benbi, D. K., & Chand, M. (2007). Quantifying the effect of soil organic matter on indigenous soil N supply and wheat productivity in semiarid sub-tropical India. *Nutrient Cycling in Agroecosystems, 79,* 103–112.

Benbi, D. K., & Senapati, N. (2010). Soil aggregation and carbon and nitrogen stabilization in relation to residue and manure application in rice-wheat systems in northwest India. *Nutrient Cycling in Agroecosystems, 87,* 233–247.

Benbi, D. K., Toor, A. S., & Kumar, S. (2012). Management of organic amendments in rice-wheat cropping system determines the pool where carbon is sequestered. *Plant and Soil, 360,* 145–162.

Bhatia, A., Pathak, H., Aggarwal, P. K., & Jain, N. (2010). Trade-off between productivity enhancement and global warming potential of rice and wheat in India. *Nutrient Cycling in Agroecosystems, 86,* 413–424.

Bhatia, A., Pathak, H., Jain, N., Singh, P. K., & Tomer, R. (2012). Greenhouse gas mitigation in rice-wheat system with leaf color chart-based urea application. *Environmental Monitoring and Assessment, 184,* 3095–3107.

Bhatia, A., Sasmal, S., Jain, N., Kumar, R., & Singh, A. (2010). Mitigating nitrous oxide emission from soil under conventional and no-tillage in wheat using nitrifi cation inhibitors. *Agriculture, Ecosystems and Environment, 136,* 247–253.

Bijay-Singh, Varinderpal-Singh, Yadvinder-Singh, Thind, H. S., Kumar, A., Gupta, R. K., et al. (2012). Fixed-time adjustable dose site-specific fertilizer nitrogen management in transplanted irrigated rice (*Oryza sativa* L.) in South Asia. *Field Crops Research, 126,* 63–69.

CAST. (2004). *Council for agricultural science and technology (CAST). Climate change and greenhouse gas mitigation: challenges and opportunities for agriculture Report 141* (p. 120).

Easterling, W. E., Aggarwal, P. K., Batima, P., Brander, K. M., Erda, L., Howden, S. M., et al. (2007). Food, fibre and forest products. In M. L. Parry, O. F. Canziani, J. P. Palutikof, P. van der Linden, & C. E. Hansen (Eds.), *Climate change 2007: Impacts, adaptation and vulnerability. Contribution of working group II to the intergovernmental panel on climate change* (pp. 307–328). Cambridge: Cambridge University Press. Fourth Assessment Report.

Goose, R. J., & Johnson, B. E. (1993). Effect of urea pellet size and dyciandiamide on residual ammonium in field microplots. *Communications in Soil Science and Plant Analysis, 24,* 397–409.

Groot Jeroen, C. J., Walter, A. H., Egbert, R., & Lantinga, A. (2006). Evolution of farm management, nitrogen efficiency and economic performance on Dutch dairy farms reducing external inputs. *Livestock Science, 100,* 99–110.

Holtz, B. A., & Caesar-Ton That, T. C. (2004). Wood chipping almond brush to reduced air polluation as a sustainable alternative to burning that enhances soil quality and microbial diverasity. In R. T. Lartey & A. J. Caesar (Eds.), *Emerging concepts in plant health management* (pp. 159–185). Kerala: Research Signpost.

Jalota, S. K., Kaur, H., Kaur, S., & Vashisht, B. B. (2013). Impact of climate change scenario on yield, water and nitrogen—balance and—use efficiency of rice-wheat cropping system. *Agricultural Water Management, 116,* 29–38.

Jalota, S. K., Kaur, H., Ray, S. S., Tripathy, R., Vashisht, B. B., & Bal, S. K. (2012). Mitigating future climate change effects by shifting planting dates of crops in rice-wheat cropping system. *Regional Environmental Change, 12,* 913–922.

Jalota, S. K., Khera, R., Arora, V. K., & Beri, V. (2007). Benefits of straw mulching in crop production. *Journal of Research (PAU), 44,* 104–107.

Jalota, S. K., & Prihar, S. S. (1998). *Reducing soil water evaporation by tillage and straw mulching.* Ames: IOWA State University Press [Chapter 5].

Jalota, S. K., & Vashisht, B. B. (2016). Adapting cropping systems to future climate change scenario in three agro-climatic zones of Punjab, India. *Journal of Agrometeorology, 18,* 48–56.

Jalota, S. K., Vashisht, B. B., Kaur, H., Arora, V. K., Vashist, K. K., & Deol, K. S. (2011). Water and Nitrogen-balance and -use efficiency in a rice (*Oryza sativa*)-wheat (*Triticumaestivum*) cropping system as influenced by management interventions: field and simulation study. *Experimental Agriculture, 47,* 609–628.

Jalota, S. K., Vashisht, B. B., Kaur, H., Kaur, S., & Kaur, P. (2014). Location specific climate change scenario and its Impact on rice-wheat cropping system. *Agricultural Systems, 131,* 77–86.

Kallenbach, C. M., Rolston, D. E., & Horwath, W. R. (2010). Cover cropping affects soil N_2O and CO_2 emissions differently depending on type of irrigation. *Agriculture, Ecosystems & Environment, 137,* 251–260.

Kukal, S. S., Rassol, R., & Benbi, D. K. (2009). Soil organic carbon sequestration in relation to organic and inorganic fertilization in rice-wheat and maize-wheat systems. *Soil and Tillage Research, 102,* 87–92.

Ladha, J. K., Pathak, H., Krupnik, T. J., Six, J., & Van Kessel, C. (2005). Efficiency of fertilizer nitrogen in cereal production: Retrospect and prospects. *Advances in Agronomy, 87,* 85–156.

Ladha, J. K., Reddy, C. K., Padre, A. T., & Van Kessel, C. (2011). Role of nitrogen fertilization in sustaining organic matter in cultivated soils. *Journal of Environmental Quality, 40,* 1756–1766.

Lal, R. (2004). Soil carbon sequestration to mitigate climate change. *Geoderma, 123,* 1–22.

Lehmann, J., Gaunt, J. L., & Rondon, M. (2006). Bio-char sequestration in terrestrial ecosystems—A review. *Mitigation and Adaptation Strategies for Global Change, 11,* 403–427.

Lu, C. Q., Tian, H. Q., Liu, M. L., Ren, W., Xu, X. F., Chen, G. S., et al. (2012). Effect of nitrogen deposition on China's terrestrial carbon uptake in the context of multifactor environmental changes. *Ecological Applications, 22,* 53–75.

Lu, F., Wang, X. K., Han, B., Ouyang, Z. Y., Duan, X. N., & Zheng, H. (2010). Net mitigation potential of straw return to Chinese cropland: Estimation with a full greenhouse gas budget model. *Ecological Applications*, *20*, 634–647.

Matthews, R. B., Rivington, M., Muhammed, S., Newton, A. C., & Hallett, P. D. (2013). Adapting crops and cropping systems to future climates to ensure food security: The role of crop modeling. *Global Food Security*, *2*, 24–28.

McTaggart, I. P., Clayton, H., & Smith, K. A. (1994). Nitrous oxide flux from fertilized grassland: Strategies for reducing emissions. In J. van Ham, L. J. H. M. Jassen, & R. J. Swart (Eds.), *Non-CO2 greenhouse gases* (pp. 421–426). Dordrecht: Kluwer.

Meyer, B. C., & Rannow, S. (2013). Landscape ecology and climate change adaptation: New perspectives in managing the change. *Regional Environmental Change*, *13*, 739–741.

Nandel, C., Kerusebaum, K. C., Mirschel, W., & Wenkel, K. O. (2012). Testing farm management options as climate change adaptation strategies using the MONICA model. *European Journal of Agronomy*, *52*, 47–56.

Neue, H. U. (1993). Methane emission from rice fields. *BioSciences*, *43*, 466–474.

Pathak, H. (2010). Mitigating greenhouse gas and nitrogen loss with improved fertiliser management in rice: quantification and economic assessment. *Nutrient Cycling in Agroecosystems*, *87*, 443–454.

Petersen, S. O., Regina, K., Pollinger, A., Rigler, E., Valli, L., Yamulki, S., et al. (2006). Nitrous oxide emissions from organic and conventional crop rotations in five European countries. *Agriculture, Ecosystems and Environment*, *112*, 200–206.

Rasmussen, P. E., & Collins, H. P. (1991). Long-term impacts of tillage, fertilizer, and crop residue on soil organic matter in temperate semiarid regions. *Advances in Agronomy*, *45*, 93–134.

Rivington, M., Matthews, K. B., Buchen, K., Miller, D. G., Bellocchi, G., & Russell, G. (2013). Climate change impacts and adaptation scope for agriculture indicated by agro-meteorological metrics. *Agricultural Systems*, *114*, 15–31.

Rodriguez-Lizana, A., Espejo-Perez, A. J., Gonzalez-Fernandez, P., & Ordonez-Fernandez, R. (2008). Pruning residues as an alternative to traditional tillage to reduce erosion and pollutant dispersion in olive groves. *Water, Air, and Soil Pollution*, *193*, 165–173.

Saha, S. K., Nair, P. K. R., Nair, V. D., & Kumar, B. M. (2010). Carbon storage in relation to soil size-fractions under some tropical tree based land used systems. *Plant and Soil*, *328*, 433–446.

Saharawat, Y. S., Ladha, J. K., Pathak, H., Gathala, M., Chaudhary, N., & Jat, M. L. (2012). Simulation of resource-conserving technologies on productivity, income and greenhouse gas emission in rice-wheat system. *Journal of Soil Science and Environmental Management*, *3*, 9–22.

Sayre, L. (2003). *Organic farming combats global warming—Big time, web article, Rodale Institute*. http://www.rodaleinstitute.org/ob31.

Shang, Q. Y., Yang, X. X., Gao, C. M., Wu, P. P., Liu, J. J., Xu, Y. C., et al. (2011). Net annual global warming potential and greenhouse gas intensity in Chinese double rice-cropping systems: A 3-year field measurement in long-term fertilizer experiments. *Global Change Biology*, *17*, 2196–2210.

Sierra, M., Martinez, F. J., Verde, R., Martin, F. J., & Macias, F. (2013). Soil-carbon sequestration and soil-carbon fractions, comparison between poplar plantations and corn crops in south-eastern Spain. *Soil and Tillage Research*, *130*(1–6).

Singh, G., Jalota, S. K., & Sidhu, B. S. (2005). Soil physical and hydraulic properties in a rice-wheat cropping system in India: Effects of rice-straw management. *Soil Use and Management*, *231*, 17–21.

Singh, G., Jalota, S. K., & Yadvinder-Singh (2007). Manuring and crop residue management effects on physical properties of soil under the rice–wheat system in Punjab (India). *Soil and Tillage Research, 94*, 229–238.

Singh, B., & Sandhu, B. S. (1978). Growth response of forage maize (*Zea mays* L.) to hydrothermal regime of soil as influenced by irrigation and mulching. *Indian Journal of Ecology, 5*, 181–191.

Skiba, U., Fowler, D., & Smith, K. A. (1997). Nitric oxide emissions from agricultural soils in temperate and tropical climates: Sources, controls and mitigation options. *Nutrient Cycling in Agroecosystems, 48*, 139–153.

Smith, P., Martino, D., Cai, Z., Gwary, D., Janzen, H., Kumar, P., et al. (2008). Greenhouse gas mitigation in agriculture. *Philolosphical Transactions Royal Society. Biological, 363*, 789–813.

Sodhi, G., Beri, V., & Benbi, D. K. (2009). Soil aggregation and distribution of carbon and nitrogen in different fractions under long-term application of compost in rice–wheat system. *Soil and Tillage Research, 103*, 412–418.

Sohi, S., Lopez-Capel, E., Krull, E., & Bol, R. (2009). *Biochar, climate change and soil: A review to guide future research.* CSIRO Report Number 05/09.

Stern, N. (2006). *The economics of climate change: The Stern review.* Cambridge: Cambridge University Press.

Thind, H. S., Bijay-Singh, Pannu, R. P. S., Yadvinder-Singh, Varinderpal-Singh, Gupta, R. K., et al. (2010). Relative performance of neem (*Azadirachta indica*) coated urea vis-a-vis ordinary urea applied to rice on the basis of soil test or following need based nitrogen management using leaf colour chart. *Nutrient Cycling in Agroecosystems, 87*, 1–8.

Topp, E. (1993). Effects of selected agrochemicals on methane oxidation by an organic agricultural soil. *Canadian Journal of Soil Science, 73*, 287–291.

Van Camp, L., Bujjarabal, B., Gentile, A. R., Jones, R. J. A., Montanarella, L., Olazabal, C., et al. (2004). *Reports of the technical working groups established under the thematic strategy for soil protection. EUR 21319 EN/1* (p. 872). Luxemburg: Office for Official Publications of the European Communities.

Vashisht, B. B., Mulla, D. J., Jalota, S. K., Kaur, S., Kaur, H., & Singh, S. (2013). Productivity of rain-fed wheat as affected by climate change scenario in northeastern Punjab, India. *Regional Environmental Change, 13*, 989–998.

Vashisht, B. B., Nigon, T., Mulla, D. J., Rosen, C., Xu, H., Twine, T., et al. (2015). Adaptation of water and nitrogen management to future climates for sustaining potato yield in Minnesota: Field and simulation study. *Agricultural Water Management, 152*, 198–206.

Venterea, R. T., Burger, M., & Spokas, K. A. (2005). Nitrogen oxide and methane emissions under varying tillage and fertilizer management. *Journal of Environmental Quality, 34*, 1467–1477.

Young, R. R., Wilson, B., Harden, S., & Bernadri, A. (2009). Accumulation of soil C under zero tillage cropping and perennial vegetation on the Liverpool Plains, eastern Australia. *Australian Journal of Soil Research, 47*, 273–285.

INDEX

Note: Page numbers followed by *f* indicate figures, *t* indicate tables, *b* indicate boxes, and *np* indicate footnotes.

Printed in the United States
By Bookmasters